GREEN BUILDING FUNDAMENTALS

A PRACTICAL GUIDE TO UNDERSTANDING AND APPLYING FUNDAMENTAL SUSTAINABLE CONSTRUCTION PRACTICES AND THE LEED® GREEN BUILDING RATING SYSTEM™

Michael Montoya, PE, LEEDAP

Prentice Hall
Upper Saddle River, New Jersey
Columbus, Ohio

Library of Congress Cataloging-in-Publication Data

Montoya, Michael.
   Green building fundamentals: a practical guide to understanding and applying fundamental
sustainable construction practices and the (LEED) green building rating system / Michael Montoya.
     p. cm.
   Includes bibliographical references and index.
   ISBN-13: 978-0-13-502839-1
   ISBN-10: 0-13-502839-6
   1. Sustainable buildings—Design and construction.   2. Leadership in Energy and Environmental
Design Green Building Rating System.   I. Title.
   TH880.M66 2010
   720′.47—dc22

                    2008034253

Vice President and Executive Publisher: Vernon R. Anthony
Acquisitions Editor: Eric Krassow
Editorial Assistant: Sonya Kottcamp
Production Coordination: Ravi Bhatt/Aptara®, Inc.
Project Manager: Maren L. Miller
AV Project Manager: Janet Portisch
Operations Specialist: Laura Weaver
Art Director: Jayne Conte
Cover Designer: Bruce Kenselaar
Cover Research and Permissions: Catherine Mazzucca
Cover Image: © 2006 Alexander Isley Inc./U.S. Green Building Council
Director of Marketing: David Gesell
Executive Marketing Manager: Derril Trakalo
Senior Marketing Coordinator: Alicia Wozniak

This book was set in Garamond by Aptara®, Inc. and was printed and bound by Hamilton Printing
Company. The cover was printed by Demand Production Center.

Pearson Prentice Hall™ is a trademark of Pearson Education, Inc.
Pearson® is a registered trademark of Pearson plc.
Prentice Hall® is a registered trademark of Pearson Education, Inc.

Pearson Education Ltd., London
Pearson Education Singapore Pte. Ltd.
Pearson Education Canada, Inc.
Pearson Education—Japan

Pearson Education Australia Pty. Limited
Pearson Education North Asia Ltd., Hong Kong
Pearson Educación de Mexico, S.A. de C.V.
Pearson Education Malaysia Pte. Ltd.

Prentice Hall
is an imprint of

www.pearsonhighered.com

10 9 8 7 6 5 4 3 2 1
ISBN-13: 978-0-13-502839-1
ISBN-10:   0-13-502839-6

This book is dedicated to my sweet little princess, Mia;
I hope it will help make the world a better place
for you to grow up in.

CONTENTS

PREFACE

Over the last few years, I have observed a fundamental change in the building industry. As global attention focuses more on climate change, more information is available to evaluate our environmental performance as designers and builders. It is clear that the building industry currently is part of the problem, not the solution. Relatively recently, methods were developed for evaluating the environmental performance of buildings and action was taken to improve it, notably the U.S. Green Building Council's (USGBC) Leadership in Energy and Environmental Design (LEED) Green Building Rating System. As the industry continues to evaluate these high-performance buildings, it is becoming clear that the environmental performance can be improved. Builders have demonstrated that it is possible to improve without significant cost implications and that we can actually save money by being more efficient while providing a healthy environment for building occupants and the surrounding community. These realizations and the exponential growth in the green building industry are causing a fundamental change in the way projects are approached. I wrote this book as a guide to help every construction professional understand the fundamental concepts associated with improving environmental performance that should be implemented on every project.

This book is intended to familiarize the reader with the fundamental sustainable design and green building practices and their implementation. The goal is to improve the environmental performance of buildings. In order for this goal to be achieved, every building professional must consider fundamental practices that can and should be applied to every project. This book is intended, not to compete with other publications, but instead to compile information from multiple sources into one place and to supplement currently available publications. It can be used as a supplemental guide to the USGBC's LEED for New Construction and Major Renovations Reference Guide. The book summarizes the fundamental concepts of sustainable design and green building practices, explains the big picture and why sustainability is important, and expands the discussion of building strategies and technologies by describing methods that are being used worldwide to improve the environmental performance of buildings. The book also includes a section on how to become a LEED Accredited Professional, which in turn includes a Focused Study Plan and a Study Assessment Guide with sample questions and tips to help focus your study. This is a practical guide for everyone involved in the building process: designers, superintendents, inspectors, planners, owners, engineers, facilities managers, managers, and CEOs.

Readers will learn how buildings can be designed more effectively by following simple guidelines, like orienting the building to take advantage of passive ventilation strategies and installing systems that maximize the efficiency of a building's water and power use. The book examines methods that are being used to improve environmental performance, such as on-site power generation and more efficient use of building materials. The book is intended to help builders understand how to improve construction practices, such as waste reduction and recycling, and to help them appreciate the magnitude of the environmental impacts from the building process. The book is not just about the environment; it is also about the business bottom line and how improving environmental performance may affect project costs and business practices.

The book is also intended for practitioners who want to develop professionally and become LEED Accredited Professionals (LEEDAPS). The Study Guide section will guide readers through the process of accreditation and will help focus their study of the implementation of the LEED Green Building Rating System. It is a practical guide to preparing for the LEEDAP Exam.

In sum, this book is for everybody involved in the building process who wants to understand:

- The fundamental concepts of sustainable design and green building practices.
- The economic, environmental, and social imperative for action to improve environmental performance.
- The concept that green building fundamental concepts are not a far divergence from traditional methods, just strategies that are slightly different from and better than traditional ways.
- How global environmental concerns are causing a fundamental shift in the way the building process is approached.
- Why everyone involved in the building process should understand how to implement and measure green building practices using quantifiable methods such as LEED.
- How to focus your professional development, improve your individual contribution to environmental performance, and become a LEEDAP.

The building industry is experiencing a fundamental change in the way it approaches projects, and it must adapt in order to improve its environmental performance. It is clear that businesses that refuse to adapt will be left behind. Perhaps historians will look back at this time as the "Environmental Revolution."

ACKNOWLEDGMENTS

This text would not have been possible without the help of many people. I thank reviewers Andrew J. Hoffman of the University of Michigan, Brett Gunnink of Montana State University, and Dr. Richard Burt of Texas A&M University—also Hal Johnston and Scott Kelting, LEEDAP of California Polytechnic State University San Luis Obispo for their technical review. I thank all of the wonderful people at Prentice Hall, as well as editors Heather Sisan and Brian Baker of Write With, Inc. I am indebted to the students in the Construction Management Program at Cal Poly who constantly challenge and encourage me and whose energy keeps me going. I truly hope this book will encourage you to strive for better environmental performance and that my contribution will make the world a better place for you to live. I thank my industry partners and supporters who consistently strive to improve environmental performance. Finally, to my friends and family: No words can express my gratitude for your support.

Michael Montoya

CHAPTER 1

Introduction to Green Building Fundamentals

Sustainability can be defined as an ability to carry forward, support, or maintain a practice or situation for a prolonged period of time, approaching perpetuity. When this concept is applied to the built environment, it can be viewed in the context of "green" or environmentally sensitive and conscious design and construction practices. Sustainable practices specifically consider a building project's long-term impact on the environment.

The environment and the health and well-being of people are significantly affected by the buildings in which we live and work. Since modern design and construction practices were introduced, the building industry has given very little thought to improving environmental performance and has taken very little action to do so. As the world becomes more and more aware of the environmental implications of human activities, it is clear that the building industry is part of the problem. It is critical for all building practitioners to understanding fundamental concepts of sustainable design and green building, and to take action to change their practices, in order to improve the industry's environmental performance.

The extraction of building materials such as wood, steel, and cement causes water and air pollution and degrades the land on which people live and play (among other things). Many of the world's beaches, rivers, and lakes are so polluted that they are unsafe for human use and consumption. Much of this pollution can be attributed directly to the built environment. Water pollution harms plants and wildlife, infiltrates groundwater, and makes this water unsuitable for drinking or for irrigating food-producing crops.

Building practices also contribute to air pollution. The traditional building design and construction methods that we commonly use pollute the air and deplete the ozone layer. Most of the world's power is supplied by burning fossil fuels, and that contributes a significant amount of pollution to the air. Interestingly, many buildings have indoor air quality that is worse than polluted outdoor air. Ventilation systems with inefficient filtration often recirculate "old" indoor air that is polluted by human use and the chemical off-gassing from building materials such as carpet and paint. The effects on human health and well-being are staggering. Most people spend a significant amount of time indoors in buildings with air quality so poor that it affects the health of building occupants. Poor indoor air quality may increase the number of sick days that workers take, exacerbate health problems such as asthma, and reduce workers' productivity. Many people are surprised to realize that the average house causes more air pollution than the average car or that breathing the air pollution in some of the world's cities is equivalent to smoking cigarettes. Green building is not just for environmentalists, but for all people who like to breathe clean air.

Buildings consume a significant amount of the world's energy. Most of the structures that we build and inhabit are highly inefficient in their energy use. For example, inefficient heating and cooling systems and poorly designed lighting systems can increase a building's energy use significantly. Energy generation is one of the greatest contributors to global pollution. Inefficient energy use also affects a building's long-term operational cost. This is a fundamental concern for all building owners as the cost of energy continues to increase worldwide.

To achieve the goal of environmental sustainability and to truly embrace sustainable building concepts, a project must consider environmental performance holistically. A holistic, integrated approach—rather than the improvement of individual systems—allows project managers to evaluate the interrelatedness of impacts. Such an approach is critical to maximizing the efficiency of the entire project.

Early in the process, project managers should define sustainability goals and facilitate a collaborative effort by all involved, in order to maximize the chances for success. Project teams must consider environmental performance from the harvesting of building materials to the building's deconstruction at the end of its useful life. Building projects have a great environmental impact on the Earth and on the health and well-being of its occupants. Sustainably designed buildings are highly efficient with minimal environmental implications. All buildings should provide a healthy and comfortable environment for building occupants without adversely affecting the surrounding community. Green building is the implementation of sustainability practices and must further consider the impact of the construction process on the environment.

These fundamental concepts of sustainability not only apply to building projects that are called "green," but also should be considered best practices and should be applied to every building ever built.

With the relatively recent advent of methods for evaluating environmental performance, the building industry is becoming an active participant in evaluating and implementing new, more environmentally sensitive practices. This combination of the realization that traditional practices are harmful and the availability of methods for evaluation is resulting in a fundamental shift in how the building industry does business. There is clearly an economic imperative for every business to take action; those who don't will be left behind.

In order to improve environmental performance, it is critical for us to understand and measure the environmental impacts of building design and construction practices. Everyone involved in the building process contributes to the environmental performance of the project. To improve environmental performance, we have to evaluate impacts from the planning stage to the operation of the building and consider how the building can be recycled at the end of its useful life (deconstruction). Decisions made at every stage of the building project can have long-term environmental implications. In order to improve environmental performance, everyone involved with a project should understand and apply fundamental sustainable design and green building practices.

Until relatively recently, the building industry has not had a method for measuring environmental performance. With the development of measurement standards to evaluate environmental performance, the building industry is quickly beginning to understand that traditional methods are inefficient and harmful to the environment. One of the most notable measurement standards is the U.S. Green Building Council's (USGBC) Leadership in Energy and Environmental Design (LEED®) Green Building Rating System™, which has achieved an unprecedented level of acceptance by the building industry. Buildings constructed to LEED standards have proven that traditional design and building practices are resulting in unacceptable environmental performance and that we can improve substantially on those practices. All levels of government, owners, designers, and builders are embracing the LEED rating system as a means for improvement. From a measurement perspective, no single method will work for everyone and the goals and methods are continuously evolving. However, the fundamental concepts and practices apply to every new project.

The building industry is experiencing a fundamental change in the way we approach projects and must adapt in order to improve environmental performance. When we are successful, historians will look back at this time as the "Environmental Revolution." As Dr. Hina Pendle wrote in her article "The Social Side of Sustainability,"

> The exponential growth we are experiencing today in the sustainability and environmental movements are literally the direct results of a "dying" era. We are living with the unanticipated consequences of having developed many technologies and methodologies that are destructive to life. It is no longer okay to "do it" or "make it" just because "we can." Many of us are looking for life-affirming ways of seeing, doing, and being that don't compromise the present or the future well-being of life on earth.

This book is intended to familiarize the reader with the fundamental sustainable design and green building practices and their implementation. As such, it is a practical guide for everyone involved in the building process: *designers, administrators, superintendents, inspectors, drafters, planners, facility managers, owners, engineers, managers, and CEOs*. In this book, readers will learn how simple things like orienting a building to take advantage of passive ventilation can improve its efficiency. The book also covers more complicated concepts, such as how to maximize the efficiency of a building's water and power use. We will look at methods that are successfully being used to improve environmental performance, such as on-site power generation, techniques to improve the quality of the indoor environment, and more efficient use of building materials. Builders who read the pages that follow will come to understand practices such as waste reduction and recycling, and to appreciate the magnitude of the environmental impacts from the building process. This book is not just about the environment; it's also about the building industry's bottom line, and it examines how improving environmental performance may affect cost.

The book is also intended for practitioners who want to develop professionally and become a LEED Accredited Professional (LEEDAP). The Study Guide section will guide the reader through the process of accreditation, help the reader understand the implementation of the LEED Green Building Rating System, and help the reader prepare for the LEEDAP Exam.

The chapters that follow will help everybody involved in the building process to understand

- The fundamental concepts of sustainable design and green building practices.
- The economic, environmental, and social imperative for action to improve environmental performance.

- The concept that green building fundamentals are not a far divergence from traditional methods—just different and better.
- The fundamental shift that global environmental concerns are causing in the way we approach the building process.
- The reasons that everyone involved in the building process should understand how to implement and measure green building practices using quantifiable methods such as LEED or an equivalent system if one is ever developed.
- The way to focus your professional development and individual environmental performance to become a LEEDAP.

REFERENCES AND RESOURCES FOR FURTHER STUDY

1. Pendle, Hina. *The Social Side of Sustainability*. Green Building Information, iGreenBuild.com, 2007.
2. "LEED" and related logo are trademarked by the U.S. Green Building Council and used by permission.
3. "LEED Green Building Rating System" is a trademark owned by the U.S. Green Building Council and is used by permission.
4. All text, graphics, layout, and other elements of content referring to the U.S. Green Building Council's Leadership in Energy and Environmental Design (LEED®) Green Building Rating System™ are protected by copyright under both United States and foreign laws. Used by permission.

Sustainability and the Building Industry

OUTLINE

- The Triple Bottom Line
- The Environmental Imperative
- The Economic Imperative
- The Social Imperative
- References and Resources for Further Study

THE TRIPLE BOTTOM LINE

It is critical for all building professionals to understand the "triple bottom line" in order for them to keep pace with the industry's demand for improved environmental performance. The increasing global awareness of the environmental impacts of building is causing a fundamental shift in the way the building industry approaches projects and is providing an imperative for action. As government agencies and private developers increase the demand for green buildings, building professionals and businesses must adapt or be left behind. Most buildings demonstrate a low level of environmental performance, but green buildings that are complete clearly show that a high level of environmental performance can be achieved when builders incorporate sustainable practices into projects. The building industry is now finding that this discussion affects not only the environment, but also the way we do business and the industry's bottom line. Awareness of the triple bottom line promotes environmental, economic, and social improvements in the way we do business and approach projects.

(A note on statistics: The statistics presented in this book are from multiple sources or original research and are intended to demonstrate an order of magnitude. Skeptics should consider that, even taking statistical variances into account, the statistics are all valid representations of an order of magnitude to support this discussion.)

THE ENVIRONMENTAL IMPERATIVE

The world is becoming more and more aware of the detrimental impacts of human-generated pollution on our environment and quality of life. The quality of the indoor environment of many of the buildings in which we live and work is so poor that it affects the health and well-being of the building occupants. Furthermore, many of these buildings were not designed to foster social interaction, community connectivity, or connections with natural spaces. Such features are all necessary to provide a healthy life for building occupants and surrounding communities.

There are a number of concerns about environmental performance with regard to how we design and build:

- Most of the energy we use in our buildings comes from burning fossil fuel at power plants, a process that causes air pollution, smog, acid rain, global warming, and other adverse environmental impacts. According to the U.S. Environmental Protection Agency (EPA), emissions from power plants contribute to over 40,000 deaths per year from lung cancer and heart attacks that are due primarily to air pollution.
- The building industry's use of natural resources and the pollution it generates have a significant effect on the environment. It is estimated that the buildings in the United States account for approximately
  - 71% of our electrical energy consumption,
  - 39% of our greenhouse gas emissions,
  - 30% of the raw materials we use (buildings worldwide account for 40% of raw materials used worldwide),
  - 30% of our waste output, and
  - 12% of our potable water usage.
- Water is a natural resource necessary to support life. The availability of clean water is quickly diminishing with pollution and increased demand due to population growth. The world's water supplies available for human consumption, irrigation for food production, and recreation are in a sad state. For example,
  - Over 45% of U.S. lakes are too polluted to allow fishing or swimming or to support aquatic life.
  - Trillions of gallons of untreated sewage, stormwater, and industrial waste are discharged into U.S. waters annually.
  - Approximately 25% of U.S. beaches are closed due to water pollution in any given year.
  - Approximately half of the world's population is affected by drinking water pollution, which annually causes 5–10 million deaths and over 250 million cases of water-related diseases such as cholera, typhoid, schistosomiasis, and dysentery.
  - Around 80% of the pollution in the world's oceans comes from land-based activities.
- Water pollution in rivers, lakes, and oceans is attributed primarily to polluted runoff from
  - Roads, parking lots, and landscaping.
  - Farms; included in such runoff are pesticides, fertilizers, and animal waste.
  - Logging and mining operations.
- Manufacturing products such as steel and aluminum from natural resources consumes a significant amount of energy. The material manufacturing process also results in significant environmental pollution. For example, the embodied energy (the amount of energy it takes to produce one pound of a product) necessary to create steel from ore is about 19,200 BTU/pound, one of the highest amounts of energy for any material manufacturing process. Manufacturing one ton of steel from ore (using the basic oxygen process) consumes about

- 3,200 pounds of ore,
- 300 pounds of limestone,
- 900 pounds of coke (made from coal),
- 80 pounds of oxygen, and
- 2,600 pounds of air.

This material manufacturing process also produces a significant amount of environmental pollution, including

- 4,600 pounds of gaseous emissions,
- 600 pounds of slag, and
- 50 pounds of dust.

Steel produced from recycled materials is about 80% more efficient

- The U.S. EPA has been measuring air quality in most U.S. cities since the 1990s. The EPA's data indicate that many U.S. cities (on the order of 20%) do not meet established minimum air quality standards. In regions such as the Midwest industrial areas and California, the level of air pollution in areas that do not meet minimum air quality standards is considerably higher than national averages (as much as 60% higher)

According to the American Cancer Society, short-term exposure to the levels of air pollution that are common in U.S. cities is associated with increased risk of cardiovascular illness and death. Long-term exposure to elevated levels of air pollution is a factor in reducing overall life expectancy on the order of 1 to 3 years.

It is clear that the building industry is a contributor to the world's environmental problems and that there is an environmental imperative for us to design and build greener buildings. We are also finding that the cost savings associated with highly efficient buildings are potentially significant.

THE ECONOMIC IMPERATIVE

Historically, people have assumed that building highly efficient buildings with lower environmental impacts is significantly more expensive than building according to traditional methods. The sustainable design and building movement has gained significant popularity among building owners, builders, designers, and government agencies in recent years. The popularity of this movement is making the technology for highly efficient buildings more affordable. We are beginning to recognize that this is not just a passing fad, and that there is an economic imperative to designing and building highly efficient buildings. Designers and builders who do not embrace this concept in their business models will soon be left behind.

The sustainable design and construction movement has been called the "next marketing boom of the new millennium." The U.S. Green Building Council's (USGBC) LEED Green Building Rating System is at the forefront of this movement. In-depth knowledge of the LEED rating system is invaluable for building professionals who wish to improve the environmental and economic performance of their buildings. This knowledge is critical for builders to develop a long-term market share. There is clear evidence that sustainable design and construction practices are here to stay and will be considered "best practices" for designers and builders within the foreseeable future. It is also clear that for the foreseeable future the

LEED Rating System will be the benchmark. This expectation is supported by the fact that many government agencies are now *requiring* builders to use more sustainable design and construction practices and to follow the LEED model for both public and private projects.

- The federal government, 22 states, and 75 local governments have made policy commitments to use or encourage LEED.
- Many jurisdictions offer fast-tract permit process for builders who commit to sustainable design and construction projects.
- Several states and local governments have pending legislation that will require all new public buildings to be LEED certified.

As the demand for sustainable building technologies continues to increase, these technologies are becoming more readily available, efficient, and affordable. Owners of highly efficient buildings are realizing significant cost savings over time, primarily through energy efficiency and savings in operating costs. Building owners and developers are increasingly choosing to certify their entire portfolios. One such company is GE Real Estate, one of the world's largest real estate investment companies, which has $72 billion in assets in 31 countries and which handles $30 billion in transactions annually. GE Real Estate announced that it will use LEED to evaluate new purchases and to benchmark and retrofit existing buildings in its portfolio.

In addition to long-term cost savings from increased energy efficiency, building owners are finding that LEED-certified buildings outperform conventional buildings in sale value, rental rates, and occupancy rates. A recent report by the New Buildings Institute (NBI) indicates that LEED-certified buildings achieve energy use savings averaging 25 to 30% better than conventional buildings. Gold- and Platinum-certified buildings average almost 50% savings in energy use. A recent report by the CoStar Group indicates that LEED-certified buildings generate an average $11.24 more in rent per square foot and have a 3.8% higher occupancy rate than conventional buildings. This study also indicates that LEED-certified buildings are selling for an average of $171 more per square foot than their conventional counterparts. The return on investment (the amount of time it takes for the long-term cost savings to pay off the initial investment) is often well within the scope of justifiable investment strategies.

It is clear that the building industry is a contributor to the world's environmental problems. It is also clear that there is an economic imperative for action to improve environmental performance and to design and build greener buildings. Green building projects bring benefits not only to their occupants, but also to the larger community.

THE SOCIAL IMPERATIVE

Many of the projects that we build consider only the economic and environmental impacts on the building users and owners. Building projects also affect the overall community and society in which they are built. Buildings define and affect community spaces where people live and interact. Buildings create not only a physical environment, but also a social environment. Social sustainability considers and promotes social interaction and cultural enrichment for building users and surrounding communities. Well-designed projects emphasize the well-being of all economic and social levels of society and respect diversity and social capital. Socially sustainable projects consider potential impacts on other humans in our global community. The application of this fundamental concept ranges from improving business operations to implementing strategies in a project design that will improve social interaction. All projects should consider the basic needs of people and communities such as happiness, health, safety, freedom, economic vitality, and dignity. The concept of social sustainability is evolving as we become more conscious of the effects building projects have on communities and societies. Trevor Hancock, one of the pioneers of the "healthy communities" movement in North America, argues that "socially sustainable development

- Meets basic needs for food, shelter, education, work, income, and safe living and working conditions.
- Is equitable, ensuring that the benefits of development are distributed fairly across society.
- Enhances, or at least does not impair, the physical, mental, and social well-being of the population.
- Promotes education, creativity, and the development of human potential for the whole population.
- Preserves our cultural and biological heritage, thus strengthening our sense of connectedness to our history and environment.
- Promotes conviviality, with people living together harmoniously and in mutual support of each other.
- Is democratic, promoting citizen participation and involvement.
- Is livable, linking 'the form of the city's public places and city dwellers' social, emotional and physical well-being.'"

The unprecedented growth of sustainable design and green building has the potential to affect the economic vitality of building occupants and surrounding communities. The increased demand for sustainable building materials, technologies, and installation is affecting communities by creating "green collar" jobs for people. Sustainable projects can help communities find greater economic prosperity by generating work in green construction and trades. In order to encourage the ongoing development of this social equity, sustainable projects must educate employees, building occupants, owners, governments, communities, and other members of the global community. This outreach can also encourage individuals, families, and communities to lead more environmentally sustainable lifestyles.

Community development and urban planning affects societies by addressing issues such as food production and availability, people's access to work and education, and the level of social interaction and participation in the community. Sustainably designed projects consider community connectivity and find ways to use spaces to encourage community and social interaction.

A sustainably designed building is highly efficient, resulting in lower operating costs over the life of the building and potentially reduced living costs for the occupants. There is a significant potential to improve social equity by applying this basic concept to low-income housing. Affordable housing is often built to the lowest possible standards on the outer edges of urban areas. Traditional low-income housing adversely affects the health and well-being of the building occupants and discourages them from actively participating in society. Sustainably designed buildings are affordable to operate and are efficient, comfortable, and well planned to provide a healthy lifestyle for building occupants. Sustainably planned projects encourage people to engage in social interaction and to maintain sustainable, healthy lifestyles by encouraging them to walk instead of drive and by providing outdoor and natural gathering spaces for people to interact socially. Locating buildings near essential services such as education facilities, public transportation, food, and health care help improve people's quality of life.

Socially sustainable development strives to break down social and economic barriers in order to strengthen communities and to improve the health and well-being of all people in this and future generations.

REFERENCES AND RESOURCES FOR FURTHER STUDY

1. Allen, Edward, and Joseph Iano. *Fundamentals of Building Construction. 4th ed*. New York: John Wiley & Sons, 2004.
2. BNP Media. *Environmental Design and Construction*. Troy, MI, 2007.
3. Building Green, Inc. *Environmental Building News*. Brattleboro, VT, December 2007.
4. CoStar Group Real Estate Information. *Commercial Real Estate and the Environment*. 2008. http://www.costar.com/partners/costar-green-study.pdf.
5. Diamond, Rick, Mike Opitz, Tom Hicks, Bill Von Neida, and Shawn Herrara. *Evaluating the Site Energy Performance of the First Generation of LEED Certified Commercial Buildings*. Washington, DC: American Council for an Energy Efficient Economy, 2006.
6. Hawken, Paul. *Blessed Unrest: How the Largest Movement in the World Came into Being and Why No One Saw It Coming*. New York: Viking, 2007.
7. Hawken, Paul, Amory Lovins, and L. Hunter Lovins. *Natural Capitalism: Creating the Next Industrial Revolution*. New York: Viking, 1999. This publication explores a future in which business and environmental interests increasingly overlap and in which businesses can better satisfy their customers' needs, increase profits, and help solve environmental problems all at the same time.
8. Kibert, Charles. *Sustainable Construction, Green Building Design and Delivery*. New Jersey: John Wiley and Sons, Inc., 2008.
9. Collins, Wendy, Lesley Stone, and Lawrence Gostin. *Understanding the Relationship Between Public Health and the Built Environment*. American Journal of Public Health, September 2005. This report summarizes the relationship between the way that our communities are built and several public health outcomes, such as level of physical activity, number of traffic accidents, level of respiratory health, and degree of mental health.
10. Montoya, Mike. *LEED in Practice—Applying the USGBC Green Building System*. Informa Center for Professional Development, 2006.
11. New Buildings Institute study. Funded by USGBC with support from the U.S. Environmental Protection Agency and accessible at https://www.usgbc.org/ShowFile.aspx?DocumentID=3930.
12. United Nations Environment Programme Sustainable Buildings & Construction Initiative. *Buildings and Climate Change: Status, Challenges and Opportunities*, 2007. This report finds that significant cuts in greenhouse gas emissions can be made using existing technologies and building materials.
13. U.S. Green Building Council. *Green Building Research Funding: An Assessment of Current Activity in the United States*. 2007. This report by the 2007 USGBC Ginsberg Sustainability Fellow tracks recent federal, state, and trade association contributions to green building research funding.
14. U.S. Environmental Protection Agency Green Building Workgroup. *Buildings and the Environment: A Statistical Summary*. December 20, 2004.
15. World Business Council for Sustainable Development. *Energy Efficiency in Buildings: Business Realities and Opportunities*, 2007.
16. Yates, Janet K. *Global Engineering and Construction*. Hoboken, NJ: John Wiley and Sons, 2007.

CHAPTER 3

Sustainable Design and Green Building

OUTLINE

- Sustainable Design Practices
  - Benefits of Sustainably Designed Projects
  - Designing High-Efficiency Buildings
- Reduce, Reuse, and Recycle Building Materials
  - Natural Resource Use Reduction
- Green Building Practices
- LEED®, Follow, or Get Out of the Way
- References and Resources for Further Study

SUSTAINABLE DESIGN PRACTICES

Sustainable design considers a building's environmental implications holistically, from the planning process to the building's deconstruction at the end of its useful life. Proper planning during the design phase is critical and must consider all of the project's environmental impacts, how they interrelate, and how to minimize impacts. A building project has the potential to affect the health and well-being of the building's occupants and the surrounding community, including its open spaces, ecosystems, plants, animals, air quality, and natural resources. With proper planning, the detrimental impacts of the project on the surrounding environment can be minimized; some buildings even bring benefits to the larger community.

Sustainably designed projects are highly efficient in their design, construction, and operation. Project teams give specific attention to the environmental impacts of each system and to their interrelationships. This process begins in the planning stage of the project when the project team considers site selection, building orientation, and other factors that affect the environmental performance of the project. The choice of building systems involved in power generation and ventilation has a great impact on the building's environmental performance. The materials used to build also have environmental impacts, from harvesting to the installation process.

Sustainable design must plan for impacts on the surrounding community. For example, a new building may contribute to increased vehicular traffic, air pollution, or loss of open space, all of which affect a community's quality of life. Sustainably designed projects should

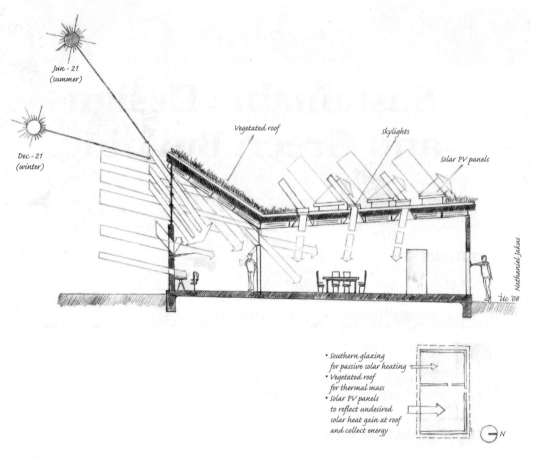

FIGURE 3.1 Sustainable building design considers natural features that are more efficient than traditional features and that improve the quality of the interior space.

not only mitigate community impacts, but actually enhance the quality of life for the building occupants and the community. People were not designed to sit in traffic breathing exhaust or to live and play on concrete and asphalt. Sustainably designed projects consider the well-being of the overall community by encouraging people to use alternate means of transportation besides driving that lessen environmental impacts. Promoting forms of transportation such as walking or riding a bike encourages people to adopt healthy lifestyles and increases the well-being of communities. The use of public transportation and carpooling is also more environmentally efficient and promotes community interaction.

With few exceptions, walking is healthier than driving—both for the environment and for the people who do it. Sustainably designed projects encourage people to tread lightly on the planet and to enjoy a healthier environment and lifestyle. A notable project that has embraced these fundamental concepts with great success is the Vauban District in Freiburg, Germany. Buildings in this district have convenient access to public transportation, and the area has natural outdoor spaces that encourage community interaction and enhance the experience of walking instead of driving. Many of the roads in this area are designed primarily

FIGURE 3.2 Signs encourage people to use environmentally efficient forms of transportation, such as carpools.

for people to walk and interact with one another by specifically giving the right-of-way to pedestrians over vehicles. It is common to see groups of people walking in the road and chatting while the occasional vehicle drives past. Signs posted throughout the development indicate that the roads are intended for walkers and that vehicles are welcome only if they travel at a walking pace.

In addition, roads between housing units in Freiburg were replaced with natural walking paths and outdoor spaces for people to interact and play together. Despite the urban setting, it is common to see children in Freiburg playing in the manmade streams, while families congregate around benches and tables in a natural environment, surrounded by trees and wildlife. Houses have ready access to convenience stores, schools, and a recreation center with a green roof that is part of the community's outdoor space.

The world's open natural spaces used for farming, food production, and recreation are diminishing due to the expansion of building projects and their associated pollution. These projects adversely affect natural ecosystems by damaging animal habitats and plant life. Many projects indiscriminately destroy natural areas or damage them more subtly by affecting their microclimates. For example, ecosystems may be affected by heat generated from road surfaces and buildings, commonly referred to as heat island effect. Sustainable design

FIGURE 3.3 Road signs in Freiburg, Germany, indicate that pedestrians have the right-of-way on the road. The policy is intended to reduce automobile use and encourage community interaction.

FIGURE 3.4   Playing soccer on the roof of the community recreation center in Freiburg, Germany.

is intended not only to mitigate this impact, but also to produce buildings that interact harmoniously with their natural surroundings and bring benefits to their occupants.

Green builders consistently recommend establishing environmental goals early in the project, ensuring that the goals are understood by all participants. Green builders also recommend selecting a project team that has a fundamental knowledge of sustainable practices and involving the builder early in the design process.

Benefits of Sustainably Designed Projects

Many buildings "force" people to drive to work in traffic, park on asphalt with only token landscape planting, and spend all day indoors with no connection to nature. There is clear and compelling evidence that this type of isolated lifestyle, removed from nature and from natural sunlight, is detrimental to a person's health and well-being. Sustainable design is intended not just to mitigate these adverse effects, but actually to enhance a person's life and experiences. There is no question that our urban lifestyle, inefficient buildings, growing population, and lack of open natural spaces make this a challenge.

Designers and planners are meeting this challenge simply by incorporating natural features into projects. Natural features help people renew their connection with nature and enjoy a closer connection to others in the community, even in the "concrete jungles"

FIGURE 3.5 An atrium can be used to improve occupants' connection to nature and exposure to natural sunlight.

of our cities. Sustainable design incorporates natural indoor spaces, such as atria and other gathering places for people, and natural building features, such as landscaped roofs.

Studies show that the average person spends over 90% of his or her time indoors. Many of the buildings that we build and occupy provide little exposure to natural daylight and limited views of the outdoors. Humans need exposure to sunlight in order to process nutrients in their bodies. Long-term lack of exposure to natural sunlight contributes to ailments such as seasonal affective disorder (SAD), which manifests with symptoms of depression. Lack of exposure to sunlight may even be related to ailments such as alcoholism. For example, there is generally a higher rate of SAD and alcoholism in arctic communities that have extended seasons with little daylight than in communities at latitudes that receive relatively steady amounts of daylight year round. Studies also show that people who live and work in buildings that do not provide views of the outdoors have a higher incidence of depression and other health problems. Just the thought of working in a windowless office or looking out a window at a neighbor's wall is depressing. Sustainable design enhances a person's experience in a building by providing natural daylighting and views. Implementing this concept is simple and it should be included in every project design. Sustainable design incorporates building features that provide exposure to natural daylighting such as atrium roofs, skylights, and strategically placed windows and exterior natural features.

Sustainable design strives to produce healthy and comfortable spaces for people not just to occupy, but actually to enjoy. Such indoor spaces are thermally comfortable for occupants and have high air quality. Highly efficient buildings maximize the efficiency and comfort of buildings by using natural means for heating, cooling, and ventilation.

FIGURE 3.6 Glass cladding systems can be used to increase natural daylighting and provide views of the outdoors; however, they have the drawback of causing solar heat gain.

Designing High-Efficiency Buildings

Energy use is a fundamental concern for all buildings. Sustainably designed buildings maximize the energy efficiency of building systems and use natural means for power generation such as solar and wind power. Designing high efficiency buildings is a critical aspect of reducing our demand for highly polluting, expensive fossil fuels.

Drinkable water is scarce throughout much of the world, and the situation will only get worse as the human population grows and as pollution of water sources increases. Principles of sustainable design strive to reduce a building's potable water use. These strategies are easy and inexpensive to implement. Every project should harvest rain, recycle used water from showers or laundry, and reuse wastewater for purposes such as flushing toilets and landscaping irrigation.

REDUCE, REUSE, AND RECYCLE BUILDING MATERIALS

Natural Resource Use Reduction

The materials we use to build our structures are diminishing our natural resources at an alarming rate. Raw materials such as wood, steel, cement, water, and oil used in our buildings are all becoming more difficult and expensive to find and harvest. The processing of these natural resources into building materials such as structural steel, concrete, lumber, and asphalt causes significant pollution and energy use. Maximizing the efficient use of our natural resources is critical to preserving the long-term health of our planet. Sustainable design

FIGURE 3.7   Highly efficient buildings produce their own sustainable power and recycle water.

principles consider the environmental impacts of the harvesting and production of building materials. These principles can be applied to all projects in order to reduce the use of natural resources and the pollution associated with their use.

One way to reduce the use of natural resources is to use rapidly renewable materials in place of traditional building materials. For example, some varieties of bamboo grow as much as a foot per day. This allows bamboo to regenerate after harvest much faster than traditional building materials such as oak or fir trees. Innovative designers are using bamboo for flooring, wall coverings, and even structural elements. Bamboo flooring can perform as well as or better than traditional materials. For centuries, bamboo has been used throughout the world for building frames and scaffolding. Studies show that some bamboo is stronger than steel in tension, which makes its use in reinforcing structural concrete intriguing. There are multiple examples of rapidly renewable materials that can provide the same level of quality and performance as traditional building materials, but that have much lower environmental impacts.

Sustainable design principles encourage the reuse and recycling of building materials in order to decrease the demand for new materials and the environmental impacts associated with their use. Examples of these practices are remodeling an existing structure instead of building a new one, reusing materials from a demolished building, and using building materials that contain recycled content.

FIGURE 3.8 Reduce, reuse, and recycle building materials to reduce natural resource use.

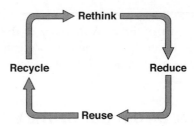

A sustainable design considers how a building will affect the environment for the duration of its useful life. A truly sustainable design considers how the building will affect the environment through its deconstruction. Such a design incorporates materials and design elements that will allow the building to be deconstructed and the materials to be reused or returned to their natural state at the end of the building's useful life.

GREEN BUILDING PRACTICES

The term "green building" has been used as a generic term to describe all the sustainability issues associated with a project. "Green building" is defined specifically as "the implementation of a sustainable design." Thoughtful practices can mitigate the significant impact of the construction process on the environment. In order to effectively implement these practices, a builder must understand the fundamental sustainable design concepts, their application to the project design, and their intent. The processes that we currently employ to construct buildings must be reevaluated to improve buildings' environmental performance. The fundamental green building processes are not more difficult to apply, just different and more efficient. Every day, builders are becoming more knowledgeable about green building practices and more efficient in implementing them. This new business paradigm is being embraced across the building industry. In order to be competitive in the new "green" world market, projects and businesses must consider fundamental practices such as

- Reducing and recycling construction waste.
- Controlling noise, light, and air pollution during construction.
- Controlling indoor air quality during and after construction.
- Protecting and restoring natural habitat.
- Limiting storm water runoff pollution and erosion.
- Ensuring that buildings operate as efficiently as possible once they are complete.
- Using materials with recycled content that are harvested and manufactured regionally.
- Choosing low-emitting building materials.
- Controlling harmful emissions from construction equipment and vehicles.

Perhaps one of the most pervasive misconceptions about green building practices is that they result in more work and increase the cost of the project. Most of the fundamental concepts of green building are not a far divergence from traditional methods, just different and more efficient. Many sustainable practices are no different in terms of the construction process. For example, low-VOC paint is applied exactly the same as traditional paint, but results in significantly less air pollution. With proactive planning and efficient implementation, it is relatively easy to incorporate green building practices into construction operations without increasing the cost of the project. Many of the builders who are considered to be the leaders in the green building movement incorporate these fundamental concepts as a matter of business, apply them to all of their projects, and offer green practices at no additional cost to the building owner.

LEED®, FOLLOW, OR GET OUT OF THE WAY

In order to meet environmental goals, the building industry must take quantifiable and measureable actions. Measuring success helps to keep us honest to our intentions and accountable for our actions. By measuring the outcome of our efforts, we can constantly evaluate and improve our operations and decrease our deleterious impacts on the environment. However, the application of fixed criteria can stifle innovation. Rather, we should develop and implement best management practices that may be tailored to every project to decrease its environmental impact. Considering and implementing fundamental environmental practices on every project allows us to expend our energy on constant innovation and improvement.

Measuring our environmental impacts is not an insignificant task. Many organizations provide thoughtful and objective methods for evaluating the environmental impacts of building projects. All these methods have merit, but they cannot be considered the only, or even the last, step in the process. These measurement tools are fundamental best practices and are only the *first* step to achieving our sustainability goals.

Of the industry's organizations, none has been as successful at achieving market penetration as the U.S. Green Building Council's (USGBC) Leadership in Energy and Environmental Design (LEED®) Green Building Rating System™. The LEED Green Building Rating System was founded to transform the way buildings and communities are designed, built, and operated with the vision of achieving sustainability within a generation. LEED is a voluntary building certification program that establishes a common standard for the measurement of what constitutes a high-performance green building. LEED gives building owners and project teams a clearly defined, practical set of design and performance goals and provides independent third-party certification that validates their achievements. In-depth knowledge of the LEED rating system not only is invaluable to builders who wish to reduce their environmental impacts, but also helps them develop a long-term market share and improves their economic performance.

- Every year U.S. buildings are responsible for 39% of U.S. CO_2 emissions and 70% of U.S. electricity consumption. Also, each year U.S. buildings use over 15 trillion gallons of water and consume 40% of the world's raw materials.
- Many conventionally built buildings have indoor air that is more polluted than the air outside. Poor indoor air quality has been linked to illnesses ranging from asthma to heart disease.
- Green buildings use an average of 30% less energy than conventional buildings, with corresponding reductions in CO_2 emissions. Achieving this level of performance in just half of the U.S. buildings constructed each year would be equivalent to taking more than one million cars off the road. Gold and Platinum LEED-certified buildings achieve average energy savings of 50%.
- Business membership in USGBC has experienced an unprecedented growth with over 15,500 member organizations and almost 100,000 active individuals as of early 2008.

FIGURE 3.9 Recently, USGBC experienced an unprecedented growth in membership.

Source: Courtesy the USGBC.

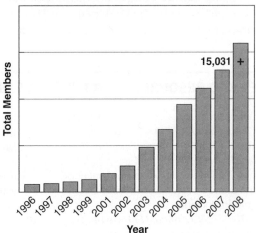

- USGBC is made up of building owners and end-users, real estate developers, facility managers, architects, designers, engineers, general contractors, subcontractors, product and building system manufacturers, government agencies, and nonprofits.
- Water conservation, reduced construction waste generation, and effective stormwater management strategies generate operational savings for the building owner while reducing demand on municipal infrastructure.
- Anecdotal studies indicate that people in green buildings have a 40–60% lower incidence of colds, flu, and asthma. Patients in green hospitals are discharged as much as two and a half days early, and kids in green schools have test scores that are up to 18% higher.
- Completed building projects have demonstrated that implementing green strategies initially costs 1–2% more than conventional building practices, but the payback period is less than a year.
- Schools designed and constructed according to LEED standards save an average $70 per square foot in operation costs, use 33% less energy, have a 40% lower occurrence of asthma, have 5–15% higher daily attendance, and have, in general, better student and teacher performance. About 75% of senior executives believe that green schools help attract and retain teachers.
- As of early 2008, over 1,500 buildings in the United States have earned LEED certification and over 11,000 more are in the certification process, representing over a billion square feet of building space.
- There are LEED projects in all 50 states and in 69 countries. Every business day, another $100 million worth of construction projects registers for LEED certification.
- Twelve federal agencies, 22 states, and over 75 local governments have made policy commitments to use or encourage the use of LEED.
- As mentioned in an earlier chapter, building owners and developers are increasingly choosing to certify their entire portfolios. For example, GE Real Estate is one of the world's largest real estate investment companies, with $72 billion in assets in 31 countries and $30 billion in transactions annually. GE intends to use LEED to evaluate new purchases and benchmark and retrofit existing buildings in the company's portfolio.

- Also mentioned in an earlier chapter, LEED-certified buildings are achieving an increase in rental rates of $11.24 per square foot and 3.8% higher occupancy rates, according to a recent report by the CoStar Group. This report also indicates that LEED-certified buildings are achieving greater sale values—on the order of $171 per square foot.

REFERENCES AND RESOURCES FOR FURTHER STUDY

1. Australia Green Building Council. *Dollars and Sense of Green Buildings 2006: Building the Business Case for Green Commercial Buildings in Australia*. This report finds that green building can reduce annual operating costs and increase return on investment, asset market value, rents, and occupant productivity.
2. BNP Media. *The Seventh Annual LEED Guide.* Troy, MI: Environmental Design and Construction, 2007.
3. BNP Media. *Environmental Design and Construction,* July 2007, Troy, MI.
4. BuildingGreen, Inc. *Environmental Building News,* December 2007, Brattleboro, VT.
5. CoStar Group Real Estate Information. *Commercial Real Estate and the Environment.* 2008. http://www.costar.com/partners/costar-green-study.pdf.
6. *GreenSource: The Magazine of Sustainable Design.* Each issue of *GreenSource* provides project case studies, information about the latest trends and technologies, a roundup of new products, and legislation and policy updates.
7. Hawken, Paul. *The Ecology of Commerce.* New York: Viking, 2004.
8. Langdon, Davis. *The Cost of Green Revisited,* 2007. This extension of the 2004 *Costing Green* report shows that many projects are achieving LEED while remaining within their budgets and in the same cost range as non-LEED projects.
9. Langdon, Davis. *Costing Green,* 2004. Based on an in-depth study of the cost of sustainable buildings, this report concludes that there is no significant difference in the construction costs for LEED and non-LEED buildings.
10. New Buildings Institute study, March 4, 2008. Funded by USGBC with support from the U.S. Environmental Protection Agency and accessible at https://www.usgbc.org/ShowFile.aspx?DocumentID=3930.
11. Normal, MacLean. *Comparing High and Low Residential Density: Life-Cycle Analysis of Energy Use and Greenhouse Gas Emissions.* Kennedy, CA: Journal of Urban Planning and Development, March 2006.
12. U.S. Green Building Council. *Sustainable Building Technical Manual (SBTM).* The SBTM is a valuable resource for designers, builders, owners, and operators who want to implement green strategies in their facilities. It is available free to USGBC members.
13. U.S. Environmental Protection Agency Green Building Workgroup. *Building and the Environment: A Statistical Summary,* 2004.

CHAPTER 4

Evaluating Cost Implications of Green Building

OUTLINE

- Cost Implications of Green Building
  - Evaluating "Soft" Costs
  - Evaluating Capital Costs and Initial Investment
- Life-Cycle Cost and Long-Term Investment Considerations
  - Selling and Renting LEED Buildings
  - Long-Term Cost Benefits Due to User Productivity
  - Building Energy and Water Consumption
- References and Resources for Further Study

COST IMPLICATIONS OF GREEN BUILDING

Investors and building owners often assume that applying green or environmentally sensitive design and construction practices will increase a project's cost. It seems intuitive that an increase in the efficiency of the building will result in an increase in the cost of building it. This assumption is being tested in over a billion square feet of green projects that have already been constructed. These projects are demonstrating that many of the fundamental concepts of green building can be applied without increasing project cost. These buildings are also demonstrating the long-term cost savings of operating a highly efficient building. For example, on-site power generation eliminates monthly electricity bills. Many potential strategies for improving environmental performance may require builders to make a larger initial investment. Built projects are demonstrating that the return on investment for many fundamental green building practices is well within the scope of common investment strategies.

There is a pervasive misconception that sustainable building solutions increase capital costs and investment risk. Investors and building owners tend to focus on short-term capital costs, chiefly because of the lack of quantifiable data on economic benefits and life-cycle cost implications of sustainability. Many people tend to see green building only in terms of its benefit to society and the environment. They view it as a short-term trend with a narrow market appeal, and no appeal at all to long-term investment planners.

We are now finding that the green buildings we have already built are achieving long-term cost benefits. Building owners are realizing quantifiable cost benefits such as

- Savings in capital cost investment.
- Reduction in operation costs.
- Increased investment returns.
- Increased occupant productivity.
- More efficient resource usage.
- Marketing and revenue generation benefits.

Based on this model, sustainable or green construction could be considered "best practices" in the design of any building.

When we quantify the costs associated with sustainability, it is important to consider the interconnectedness of different systems and to approach a project in a holistic manner. Interconnected features such as site and building design, efficient use of energy and water, resource-efficient construction, lighting and mechanical design, and building ecology should be optimized in an integrated design. For example, increasing the ventilation effectiveness of a building may require the builder to make a higher initial investment in operable windows. However, it may reduce the overall project cost due to decreased heating and cooling costs. In addition to taking this synergistic approach to sustainability, an analysis of costs should consider the long-term benefits to the building owner. Strategies must be considered and evaluated in terms of cost, marketability and increased revenue generation for rental space. While this discussion is based primarily on the construction cost considerations associated with sustainability, cost should also be considered in terms of the more holistic dynamic that affects building occupants and the environment in general.

In a typical rented tenant space, it is often assumed that a low initial cost of construction will increase the amount of revenue generated from tenant rent and decrease the amount of time needed to pay off the construction cost. This approach may preclude the use of some sustainable design and construction methods that increase the design and construction cost beyond the initial budget. Because both development and government construction projects typically have strict fiscal limitations, this is a common problem. It is the purpose of this exercise to determine whether a project can incorporate sustainable construction and design methodologies that

- Do not increase the initial cost of construction.
- Have a life-cycle cost savings that will offset the initial investment in an acceptable amount of time for investors.

The project approach must apply synergistic design concepts to the whole building and site in order to maximize the chances of success. Every project must accept an increased initial investment in sustainable, highly efficient projects with the potential for long-term savings. The initial and long-term costs are easiest to analyze in a less holistic approach during the planning phase. Sustainable design concepts have been contemplated by owners and designers throughout history. However, until the introduction of the USGBC's LEED rating system, we had no comprehensive measuring and cost tracking system. The LEED rating system is relatively young, so very little comprehensive life-cycle and building cost data have been published. One of the most comprehensive cost analyses of sustainable construction was recently published by the U.S. General Services Administration (GSA). This report compares the historical cost of constructing buildings with the cost of constructing similar buildings that incorporate sustainable methodologies based on the LEED system.

The GSA report can be a useful tool for building owners and designers to analyze the potential costs during the planning phase of a project. The results of the report are, of course, specific to the location of the buildings and to other constraints that may or may not be present in other projects. Thus, the report can be a useful tool for planning, but its conclusions must be considered in the specific context of any project. One basic assumption of the GSA report is that the buildings are located in Washington, DC, where the cost of construction averages around 5% lower than the national average. The GSA report indicates that the cost of completing a building that is LEED certified is on the order of 1% greater than the cost of completing a traditional building. A LEED Gold level certification was reported to indicate that the cost would be on the order of 8% greater. Many of the sustainable methodologies necessary to obtain a LEED certification can be incorporated into a project without increasing the overall project cost. The referenced GSA report can be found through http://www.wbdg.org/index.php.

Evaluating "Soft" Costs

A concern of many owners is that incorporating sustainable methodologies into a project will increase the design and management or "soft" costs of a project. Sustainable design practice is not a far divergence from traditional methods in terms of management and process; rather, it is a practice that is focused more on performance. Design and construction teams with a fundamental knowledge of sustainability can implement strategies that improve environmental performance as a matter of practice. In order to measure and document environmental performance, many projects choose to obtain a LEED certification, which generally requires additional monitoring and documentation over what is normally required. Documentation for the certification process may increase the project's soft costs if the project team is not knowledgeable about the process and efficient in its implementation of the additional monitoring. Many of the world's premier builders have integrated sustainable practices into their business model and have become efficient at implementing and documenting sustainable projects. Documenting sustainable performance is not a far divergence for the documentation we must all do just to avoid claims, just a little different. Builders and designers who have become efficient at this process are offering their clients LEED-certified projects at no additional soft cost.

While obtaining a LEED certification is a useful tool to incorporate sustainable design methodologies and makes the building more marketable, it is not a necessary component of sustainable design. Many owners and design teams can incorporate sustainable design methods as a matter of practice. By incorporating sustainable building practices into initial project requirements and by prequalifying design teams with experience in sustainable design methodologies, a building owner can increase the sustainability of a project without incurring additional soft costs. One way to do this is to retain designers and builders who are familiar with sustainable concepts and are efficient at implementing them. Project teams should define environmental goals early in the process and implement them as a matter of practice. The design and management activities associated with green building are not a far divergence from the methods that we traditionally use, and they do not necessarily result in additional work if the project team is efficient. If the design and management process is not significantly different from how we traditionally do business, then why would it result in additional cost?

Green builders consistently recommend making sure that all members of the project team have a fundamental knowledge of sustainable practices, thereby establishing environmental goals early in the project, ensuring that all participants understand the goals, and getting the builder involved early in the design process to help evaluate costs and improve the process.

Evaluating Capital Costs and Initial Investment

The capital costs of a project are substantially affected by the decisions made at the inception of the project. Owners and designers must balance budgetary constraints during the construction phase (i.e., capital costs), with the building's daily performance and long-term costs. Generally, a more efficient building will cost more initially. However, increased capital costs in one area may be offset by cost benefits elsewhere. For example, an analysis that considers the initial cost of a passive ventilation system must also include the potential cost savings in decreased HVAC needs. The dynamic cost relations between different systems become obvious when these systems are compared in a holistic manner.

Many strategies for increasing a building's environmental performance—such as how it is oriented on the site or whether it uses passive ventilation—have no effect on the project cost. Highly efficient building systems such as photovoltaic power generation and heating and cooling generally result in an increase in the initial cost of the project over the cost of traditional systems. This cost increase is often offset by government incentives and long-term cost savings from more efficient building performance. Most sustainable building materials are currently more expensive than traditional products and can increase the initial cost of the project. Involving the builder early in the material selection process is critical to minimizing the project cost. As the demand for sustainable building materials continues to increase, their cost will become comparable to that of traditional materials. Many sustainable products, such as bamboo flooring, have already reached parity with traditional materials. Several years ago, most builders had never heard of bamboo flooring. Now it is available at the Home Depot at a price that is comparable with other products of similar quality.

The costs of sustainable projects are constantly evolving and changing with consumer demand. Since most projects don't disclose project cost details, it is difficult to put a price tag on sustainable building. Project cost varies greatly from project to project, depending on the type and use of the building, environmental goals, and the availability of materials. There is every indication that, with current technology and materials, we can implement fundamental sustainable practices without affecting the project cost. Every project must be evaluated independently to improve environmental performance and decrease cost. According to the USGBC, on the basis of data from projects that have achieved LEED certification, the average increased initial cost is on the order of 1–2%, with a payback period of less than one year. The David and Lucile Packard Foundation report *Building for Sustainability*, published in 2002, includes a comparison of different strategies considered for each level of LEED certification for a hypothetical 9,000-square-foot office building in Los Altos, California. The report also evaluates, among other things, the energy necessary to operate the building, the building's reliance on the grid, pollution from the building's operation, external societal costs, schedule implications, and short- and long-term cost considerations. Each of these factors is evaluated and compared for each level of LEED certification. This report is a good example of the holistic approach to evaluating strategies and their cost impacts. The results indicate that, in 2002, a LEED-certified building represented about a 2% increase in construction costs compared with a traditional building. In 2004, the Davis Langdon team published an updated evaluation to look at industry developments as sustainable design became more widely accepted and used. The 2004 report titled *Costing Green: A Comprehensive Cost Database and Budget Methodology* looked at different types of buildings and concluded that there is no significant difference in average costs for green buildings compared with traditional buildings. Green building costs were further analyzed in the 2007 report *The Cost of Green Revisited*, which came to the same conclusion. According to the most recent evaluation,

> Many project teams are building green buildings with little or no added cost, and with budgets well within the cost range of non-green buildings with similar programs. We have also found that, in many areas of the country, the contracting community has embraced sustainable design, and no longer sees sustainable design requirements as additional burdens to be priced in their bids. Data from this study shows that many projects are achieving certification through [the] pursuit of the same lower cost strategies, and that more advanced, or more expensive strategies are often avoided. Most notably, few projects attempt to reach higher levels of energy reduction beyond what is required by local ordinances, or beyond what can be achieved with a minimum of cost impact.

This report is well documented with reasonable assumptions and is the most comprehensive cost evaluation for green building currently published. (View all of the preceding reports at http://www.davislangdon.com/USA/Research/.)

While the initial investment in a more efficient building must remain within the owner's budgetary constraints, it should also be considered in terms of the long-term savings realized in operating costs, or the life-cycle costs, achieved from increased building efficiencies.

LIFE-CYCLE COST AND LONG-TERM INVESTMENT CONSIDERATIONS

The long-term costs of owning and operating a building, or the life-cycle costs, are of significant concern to a building owner. Some buildings and site systems may have higher initial costs but decreased life-cycle costs. Of course, most building owners are concerned with how long the life-cycle cost savings will take to pay off the initial investment. This concept may be explored in systems such as buildings with on-site wind power generation that likely have an increased initial cost (compared with the cost of a building that relies on traditional grid power distribution) versus the long-term cost savings from the elimination of power costs to operate the building. This basic value engineering exercise can, and should be, implemented on every project.

Generally, the more efficient a building is the less it will cost to operate. Many sustainable practices that do not increase project cost bring long-term saving in the operation of the building. For example, passive ventilation strategies do not necessarily increase project cost and likely decrease the long-term operating cost, due to a reduced reliance on mechanical ventilation systems. It is often necessary for builders to make an initial investment to increase environmental performance for the sake of decreasing operating cost. For example, solar power systems are generally more expensive initially than simply connecting to the power grid. However, the power supplied by the solar panels is free, saving the cost of grid-supplied power for the life of the system.

As technology tries to keep pace with demand, more sustainable strategies become available and the price of implementing these strategies continues to improve. For example, the solar power industry has seen unprecedented growth in the last few years, and the U.S. government's goal is to achieve grid parity (the point at which solar power is less expensive than municipally supplied power) by 2015. This is clear evidence that there are long-term cost benefits to owning and operating a highly efficient building. A recent report by the NBI indicates that LEED-certified buildings can achieve up to 50% cost savings associated with decreased energy use, compared with traditional buildings.

Selling and Renting LEED Buildings

Building owners are finding that, in addition to providing long-term cost savings from increased energy efficiency, LEED-certified buildings outperform conventional buildings in sale value, rental rates, and occupancy rates. A recent report by the CoStar Group indicates that LEED-certified buildings average $11.24 per square foot higher rent generation and a 3.8% higher occupancy rate when compared to conventional buildings. This study also indicates that LEED-certified buildings are selling for an average of $171 more per square foot than their peers. The report analyzed CoStar's 44 billion square feet of commercial building space, of which about 350 million square feet are either LEED or Energy Star certified, and compared these buildings with traditional buildings of similar size, location, class, tenancy, and age.

The life-cycle costs of any project are specific to each building and are based on the actual constraints of the project, the site location, local jurisdictional regulations, and so on. Accordingly, a building owner and design team must consider the long-term cost implications of making an initial capital investment in a more efficient system.

Long-Term Cost Benefits Due to User Productivity

In addition to providing potential long-term cost savings, incorporating energy efficiency into a building may have the added benefit of increasing worker productivity among the building users. Even a small increase in productivity of the building's inhabitants can have a substantial effect on the net operating income of a building. Tenant worker productivity in terms of production rates, quality of production, and absenteeism can be affected by improving thermal comfort and setting lighting levels to reduce eye stress.

According to one national survey of large buildings, electricity typically costs on the order of $2.09 per gross square foot (adjusted for inflation, assuming a yearly inflation rate of 3%) and accounts for 85% of the building's total energy bill. The average annual cost of the electricity used by an office worker is on the order of $203 per square foot (based on national average salaries and an average of 268 square foot per office worker), which is 72 times higher than the building's energy cost. On the basis of this relationship, an increase in worker productivity of 1% will nearly offset the tenant's entire annual energy cost. While the costs of energy and the effects of energy-efficient building systems vary widely for each individual application, it is clear that monetary benefits in terms of worker productivity can be obtained by increasing a building's energy efficiency.

The Lockheed Building 157 in Sunnyvale, California, is an example of the effects of sustainable design on worker productivity. In this case study, implementing sustainability principles reduced worker absenteeism by 15%, which recouped the initial investment in less than one year. The building owner also reported that on the first major contract completed in the new building, the company realized a 15% increase in worker productivity. This increase in productivity in subsequent large projects paid for the entire building construction cost.

The West Bend Mutual Insurance Company's headquarters building in Wisconsin is another case study documenting productivity increases due to sustainable design. The building designers incorporated energy-saving features such as energy-efficient lighting, upgraded windows, and a more efficient HVAC system. The building includes "environmentally responsive" workstations that allow individual employees to control temperature, airflow, and lighting at their stations. The building also includes motion sensors that turn the workstations off when they are not in use. The company compared worker productivity between the old and new headquarters buildings and noted a productivity increase of 16% in the new building.

Following are some other documented cases of increases in worker productivity, decreases in absenteeism, and even increased sales in green buildings:

- Day lighting in Wal-Mart's "Eco-Mart" sustainable building in Lawrence, Kansas, resulted in "significantly higher" sales than in similar spaces that use traditional lighting.
- ING Bank's headquarters in Amsterdam utilized energy-efficient building systems that decreased absenteeism by 15%.
- The main post office in Reno, Nevada, conducted a lighting retrofit that had a six-year payback period and increased productivity by 6% (making a profit over the cost of the retrofit).
- Hyde Tool's lighting retrofit had a one-year payback and an increase in productivity estimated to be worth $25,000 annually.
- Pennsylvania Power and Light's lighting system upgrade produced 69% energy savings, a 13% increase in productivity, and a 25% decrease in absenteeism.

While it is intuitive that worker productivity will increase in more efficient buildings, there is little conclusive and comprehensive research available to inform the owner's and designer's decision-making process. However, these case studies provide compelling evidence that the economic benefits of sustainable design may be significantly greater than just the energy cost savings.

Building Energy and Water Consumption

A reduction in the use of energy and water and the associated long-term fiscal benefits are possibly the most widely studied aspects of sustainable building. Evidence on this topic is available to guide the decision-making process of owners and designers. Initial construction costs and long-term financial benefits are relatively easy to quantify. As with any building system, the owner and designer must compare the cost of initial construction with the long-term benefits. Photovoltaic power generation and increased efficiency lighting are two commonly used sustainable methods with relatively short payback periods. New technology is constantly emerging to make building systems more efficient while minimizing the initial cost premium. It is estimated that green buildings use an average of 36% less energy than conventional buildings and that they have corresponding reductions in CO_2 emissions.

The Lockheed Building 157 in Sunnyvale, California, is an example of an energy-efficient building. The building's architect, Leo A. Daly, incorporated design elements such as increased window sizes in order to maximize the amount of natural lighting, a central building atrium with a glazed roof, interior "light shelves" that reflect natural light within the building interior, and open office layouts. These features added two million dollars to the initial construction cost, but generated a 75% savings on lighting bills, reduced the air-conditioning load (natural light produces less heat than office lighting), and provided a 50% reduction in the overall building energy costs. The initial investment was paid back in the first four years. Increasing a building's energy efficiency can significantly increase long-term economic returns. For example, the typically assumed payback period of three years for lighting system upgrades is associated with an internal rate of return in excess of 30%, which is well within the justifiable range for most building owners.

REFERENCES AND RESOURCES FOR FURTHER STUDY

1. BNP Media. *The Seventh Annual LEED Guide*. Troy, MI: Environmental Design and Construction, 2007.

2. Capital E. *Greening America's Schools: Costs and Benefits*, 2007. This report compares the financial costs and benefits of green schools with those of conventional schools.

3. Chao, Mark, and Gretchen Parker. *Recognition of Energy Costs and Energy Performance in Commercial Property Valuation*, 2000.

4. *Green Building Project Planning and Cost Estimating*. Kingston, MA: Reed Construction Data, 2008.

5. Heschong, Lisa, et al. *Skylighting and Retail Sales: An Investigation into the Relationship Between Daylighting and Human Performance*. Heschong Mahone Group, 1999.

6. Kats, Greg. *The Costs and Financial Benefits of Green Buildings: A Report to California's Sustainable Building Task Force*, 2003.

7. Langdon, Davis. *The Cost and Benefit of Achieving Green Buildings*. 2007. This publication discusses Australia's Green Building Council Green Star Rating System, the rising market demand for high performance buildings, and the environmental and economic benefits of green building.

8. Langdon, Davis. *The Cost of Green Revisited*. 2007. This extension of the 2004 *Costing Green* report shows that many projects are achieving LEED while remaining within their budgets and in the same cost range as non-LEED projects.

9. Langdon, Davis. *Costing Green*, 2004. On the basis of an in-depth study of the cost of sustainable buildings, this report concludes that there is no significant difference in the construction costs for LEED and non-LEED buildings.

10. Lucuik, Mark, et al. *A Business Case for Green Buildings in Canada*, 2005. This compilation of recent North American building studies demonstrates that the greener the building, the higher is the net present value.

11. Montoya, Mike. *LEED in Practice—Applying the USGBC Green Building System*, Informa Center for Professional Development, 2006.

12. National Institute of Building Sciences. *WBDG: The Whole Building Design Guide*. http://www.wbdg.org/index.php. This web site provides design guidance for different types of buildings in an effort to help architects, engineers, and project managers improve the performance and quality of their buildings.

13. Portland Energy Office. *Green City Buildings: Applying the LEED Rating System*, 2000. A cost/benefit analysis of applying LEED design criteria to city buildings.

14. RREEF. *The Greening of U.S. Investment Real Estate—Market Fundamentals, Prospects and Opportunities*, 2007. This paper explores why the U.S. institutional investment real estate sector is likely to embrace sustainable building principles. It also documents trends in green building and focuses on the key drivers of green building investment, as well as the barriers that have limited this investment up to now.

15. Sustainable Building Task Force (California). *The Costs and Financial Benefits of Green Buildings*. October 2003. This report includes LEED building analyses.

16. U.S. General Services Administration. *GSA LEED Cost Study*, 2006. The GSA commissioned this report to estimate soft and hard costs for developing green federal facilities.

17. Wilson, Alex, Jenifer L. Uncapher, Lisa A. McManigal, L. Hunter Lovins, Maureen Cureton, and William D. Browning. *Green Developments: Integrating Ecology and Real Estate*, 2006. This publication describes proven procedures, potential pitfalls, and practical lessons that will help you develop environmentally sound and financially rewarding green real estate.

CHAPTER 5

Site Development Considerations

OUTLINE

- Evaluating the Project Site
- Developing Damaged Sites
  - Brownfield Site Remediation Strategies
    - Air Sparging
    - Bio Plugs
    - Bioremediation
    - Bioventing
    - Electrokinetic Remediation
    - Soil Flushing
    - Chemical Oxidation
    - Phytoremediation
    - Solar Detox
- Encouraging Alternative Transportation Use
- Reducing Disturbance to Natural Ecosystems
- Reducing Heat Island Effects on Ecosystems
- Reducing Pollution from Building and Site Lighting
- Maximizing Efficiency by Using Building Orientation
- References and Resources for Further Study

EVALUATING THE PROJECT SITE

The project site is generally selected by the owner and design team during the predesign phase. The site selection can significantly affect the environmental impact of a project. Ideally, in order to minimize the environmental impact of a project, a building site should not affect land that

- Is usable farmland.
- Is subject to flooding.
- Provides a habitat for threatened or endangered species.

FIGURE 5.1   Do not build on usable farmland.

- Is near or includes bodies of water or wetlands (areas containing plants that require or can tolerate saturated soil).
- Available as a public park or an open space.

Endangered species are plants, animals, and other organisms that are in danger of becoming extinct due to harmful human activities or environmental factors. Threatened species are likely to become endangered in the foreseeable future.

To decrease the environmental impact of a new building project, a developer should consider renovating an existing building instead. Previously developed sites are defined as having been graded or altered from their natural state. If a new building must be built, the environmental impact can be lessened if the construction takes place on a site in an urban area with existing infrastructure. Choosing a previously developed site reduces the costs and environmental impacts of a project because infrastructure such as underground utilities is already in place. In addition, siting a project in an urban area allows the future occupants of the building to work closer to where they live and reduces the environmental impact of commuter automobile use. Building occupants may have a shorter commute, may be able to take public transportation to work, or may be able to ride a bike or walk to work. Choosing an urban building site not only reduces the environmental impact from automobile use but also may improve the quality of life of building occupants. The reuse of a previously developed site also decreases the impact of human activities on open space, animal habitat, and natural resources, because undeveloped land can be preserved.

Primary site selection recommendations are to

- Construct the project on a previously developed site.
- Choose a site that is within an existing urban community.
- Choose a site close to residential areas and neighborhoods.
- Ensure that basic services (such as schools, restaurants, libraries, and medical services) are nearby.
- Provide pedestrian access to basic services for the building's occupants.

Formulas and examples for evaluating a site are included in the guide *Green Building Rating System for New Construction & Major Renovations,* Version 2.2, available from USGBC.

DEVELOPING DAMAGED SITES

Brownfield sites are properties where hazardous substances, pollution, or contaminants are present or potentially present. When developers select a brownfield site for redevelopment, they remove or contain the contaminants prior to beginning building construction, thereby reducing the environmental impact of these substances. Typically, developers are reluctant to consider brownfield sites for development, due to the cost of remediation and the potential liabilities inherent in using this land. We are now finding that these impediments are mitigated by the potential cost savings of developing brownfield sites. Consider the following examples.

- The Federal government has enacted laws to limit the liability of brownfield developers, such as the *Small Business Liability Relief and Brownfield Revitalization Act* (Public Law 107–118; H.R. 2869).
- Many states and local governments offer tax incentives to developers who remediate brownfield sites.
- Often, land that has been designated as a brownfield site may be purchased for a price below market value, offsetting the cost of remediation.
- Communities significantly benefit from brownfield site cleanup, so they may lend their support to a proposed project.

Developers should take the following steps when they are evaluating cleanup options:

- Determine the appropriate and feasible level of cleanup, and find out whether there are federal, state, or local requirements for cleanup.
- Evaluate how long the cleanup will take, how much it will cost, and what the short-term and long-term effects of the cleanup will be.
- Evaluate community concerns about protection of residents' health during cleanup and reuse of the site.
- Evaluate remediation technologies that provide the best result while minimizing any associated environmental impacts.
- Determine methods to monitor the performance and results of the cleanup.

Brownfield Site Remediation Strategies

Any brownfield site that is being considered for building should be thoroughly tested, and proposed remediation strategies should be evaluated by a licensed hazardous material professional. Generally, the following strategies are recommended:

- Choose a remediation technique that minimizes the impact of cleanup on natural features.
- Treat contaminants in place rather than off-site whenever possible.
- Monitor the site after remediation to ensure that complete contaminant removal has occurred.

This activity occurs during the predesign phase and is most commonly coordinated by the civil engineer assigned to the project. Planning for the activity occurs early in the process. However, it is imperative that owners and designers involve the builder in this process to help evaluate the practicality and cost of different strategies that are being considered. There are many technologies available for remediation of contaminants, from simply removing the material and hauling it to a disposal area to treating the material in a high-tech bioreactor. A qualified and licensed contractor should perform remediation activities, as most jurisdictions require specific licenses and permits. There are a number of commercially available remediation solutions.

AIR SPARGING.   This remediation technology removes volatile organic contaminants from soil and groundwater. Contaminant-free air or oxygen is injected into the subsurface saturated soil. The process causes liquid hydrocarbons to vaporize, whereupon they are vented.

BIO PLUGS.   This solution uses a type of bioreactor that is installed on-site to remediate contaminated groundwater. A bioreactor is a system that uses microorganisms to degrade contaminants in groundwater and soil. The microorganisms break down contaminants by

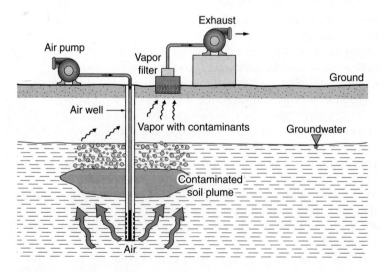

FIGURE 5.2   Air sparging is an on-site method for contaminated soil remediation.

using them as a food source. Contaminated groundwater is circulated in an aeration basin where microbes degrade organic matter, forming a sludge that is discarded or recycled.

BIOREMEDIATION.   Bioremediation uses microorganisms to degrade organic contaminants in excavated or in situ soil, sludge, and solids. The process breaks down the contaminants into gases like carbon dioxide, water, methane, hydrogen, sulfide, sulfur, and dinitrogen gas. Depending on the severity and type of soil pollution, the gas by-products may cause air pollution. If gas by-products are hazardous to people or the environment, they must be appropriately contained and discarded or recycled.

BIOVENTING.   This common form of in situ bioremediation uses extraction wells to circulate air through the ground, sometimes pumping air directly into the ground and allowing the hydrocarbon vapor to vent at the surface.

ELECTROKINETIC REMEDIATION.   This process applies a low-voltage electric current to contaminated soil to remove contaminants. The principle of electrokinetics remediation is similar to a battery. After electrodes are introduced and charged, the electric current mobilizes particles. Ions and water move toward the electrodes and can be removed.

SOIL FLUSHING.   This method injects water containing solvents into the ground at a contaminated area. The injected water, carrying contaminants, is then removed.

CHEMICAL OXIDATION.   This method converts hazardous contaminants to nonhazardous or less toxic compounds that are more stable, less mobile, and/or inert. Oxidizing agents such as ozone, hydrogen peroxide, hypochlorite, chlorine, and chlorine dioxide are introduced into the contaminated area to chemically treat the contamination.

PHYTOREMEDIATION.   This method uses plants to remove contaminants from the ground. The plants absorb contaminants through their root systems and store them in their roots, stems, and/or leaves. After the plants are harvested, fewer contaminants remain in the ground.

SOLAR DETOX.   This method is an effective way to treat contaminants after they have been removed from the ground. A catalyst is mixed with the contaminants, which are then exposed to sunlight. The ultraviolet light reacts with the catalyst, breaking down the contaminants into nontoxic substances such as carbon dioxide and water.

More information on these and other types of remediation technologies can be found in the U.S. EPA's *Road Map to Understanding Innovative Technology Options for Brownfield Investigation and Cleanup* at http://www.clu-in.org/download/misc/roadmap4.pdf.

ENCOURAGING ALTERNATIVE TRANSPORTATION USE

Automobile use significantly affects the environment. Building projects can be designed to encourage occupants to use alternate transportation means that have a lower environmental impact. Automobile use is directly associated with significant pollutants:

- Vehicle exhaust from combustion engines emits greenhouse gases and other air pollutants.

- Vehicle exhaust contributes to smog and ground level ozone that are harmful to human health.
- Motor gasoline contributes as much as 60% of carbon dioxide emissions, making it a major contributor of greenhouse gases that endanger the climate and the health of the planet.
- Producing crude oil for gasoline has significant detrimental environmental impacts.

Reducing the use of private automobiles saves energy and reduces the environmental impacts of driving. To help reduce pollution from automobile use, building projects should incorporate strategies that decrease the building occupants' reliance on automobiles for transportation. To be effective, this strategy must be considered early in the process, during the design phase. The USGBC recommends several strategies for reducing vehicle use:

- Locate the project near a commuter rail, subway station, and/or bus line to encourage building occupants to commute by mass transit, which is more efficient than individual vehicle use. The use of public transportation reduces the overall emissions generated.
- Provide on-site bicycle storage racks and shower/changing facilities in the building to encourage building occupants to commute by bicycle, which does not contribute to pollution.

FIGURE 5.3 Provide fueling stations for alternative fuel vehicles.

- Provide preferred parking and/or refueling stations for low-emission and fuel-efficient vehicles. This strategy encourages occupants to use vehicles that are highly efficient and lessens the amount of pollution generated. These vehicles include electric cars, hybrids (cars that use a combination of electricity and gasoline) and biodiesel vehicles (cars that use fuel made of organic byproducts such as cooking oil).
- Provide only a minimum amount of on-site parking, including preferred parking for carpools.

REDUCING DISTURBANCE TO NATURAL ECOSYSTEMS

Developing a site that has not been previously disturbed from its natural state, referred to as a greenfield site, disrupts the ecosystem by removing trees, other plants, and soil. Site development also destroys plant life and animal habitat. In order to mitigate this ecological damage, sites should be designed to

- Preserve the natural site elements as much as possible.
- Incorporate native plant landscaping.
- Minimize the building and hardscape (such as parking lot) areas.
- Minimize site grading to limit disturbance.
- Provide open spaces that serve as habitat and promote natural biodiversity.

During the construction phase of a project, the site disturbance can be mitigated by using readily available techniques. For example,

- Fence off areas that do not require development, in order to decrease the chances of damage to trees, other plants, water systems, and other naturally occurring elements of the location.
- Designate specific areas for site storage, disposal of waste, parking, and other construction-related activities, in order to contain disturbance.
- Fence off the trees that are to remain at the edge of their drip line (approximately the same diameter as the canopy), in order to protect them from damage and contamination from construction activities.

REDUCING HEAT ISLAND EFFECTS ON ECOSYSTEMS

Site features such as asphalt paving and roofs can absorb and radiate heat from the sun. This heat may change the natural climate of the site, potentially damaging microclimates and adversely affecting plants, animals, and humans. The amount of solar heat that a surface can reflect is measured using a Solar Reflectance Index (SRI). The SRI ranges from 0 for a black surface to 100 for a white one. Elevated temperatures can increase peak energy demand, air-conditioning costs, air pollution levels, and heat-related illness. The U.S. EPA's Heat Island Effect Initiative has published research results and potential strategies for mitigating heat islands. These documents can be found at http://www.epa.gov/hiri/.

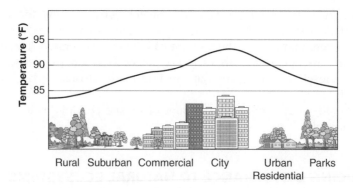

FIGURE 5.4 Urban heat island effects can increase energy use, air-conditioning costs, and air pollution.

Some of the many methods available to reduce the heat island effect include

- Shading constructed surfaces on the site with landscape features like trees.
- Using high-reflectance materials for hardscapes and roofs to reduce heat absorption.
- Replacing surfaces such as roofs, roads, and sidewalks with vegetated surfaces such as green roofs and open grid paving.

FIGURE 5.5 Green roofs have many environmental advantages.

REDUCING POLLUTION FROM BUILDING AND SITE LIGHTING

The lights installed to illuminate a building and site should be designed to prevent the light from escaping the development area. Excess light, referred to as light pollution, can

- Cause "sky glow," which impairs our view of the night sky.
- Produce glare that affects nighttime vision in motorists and pedestrians.
- Have a deleterious effect on plants and animals.

The following are methods to avoid light pollution:

- Select interior building lights that provide only the necessary light for the use of a space. Use natural sunlight as much as possible by maximizing windows in appropriate areas and installing skylights.
- Aim light fixtures to illuminate nonreflective surfaces such as opaque finishes, in order to prevent light from reflecting from its intended area of luminance and escaping through windows.
- Install shelves made of a reflective surface, often referred to as "light shelves," over windows on perimeter walls. These shelves help reflect errant light and keep it from escaping out of a window to the exterior.
- Install interior building light controllers that automatically turn off light fixtures when a space is not being used. (This strategy will also help with energy consumption.)
- For exterior lighting, use the minimum amount of light necessary for safety and comfort, and install timers or motion sensors that turn lights off when they are not needed.
- Install covers on site lights that do not allow light to project upwards or outside the project boundaries. Avoid light projection onto reflective surfaces.

The Illuminating Engineering Society of America (IESNA) provides specific lighting standards and requirements for most building types. See http://www.iesna.com for more information.

MAXIMIZING EFFICIENCY BY USING BUILDING ORIENTATION

The way that a building is oriented on a site may have one of the greatest effects on the efficiency of the building. Natural factors like wind, sunlight, and humidity can greatly affect a building's performance. Natural methods can be used to improve the efficiency of thermal and lighting systems and the comfort of a building's inhabitants. The choice of building orientation is generally made by the architect during the project's schematic design phase. To maximize the efficiency of a building's orientation, the design should consider the following factors and strategies:

- A building should be oriented to take advantage of passive and active solar strategies. For example, in a cold climate, a building should be oriented to maximize the southern exposure, allowing the sun to naturally heat interior spaces and increase the building's heating efficiency. Dark-colored exterior surfaces and upgraded insulation also help retain heat.

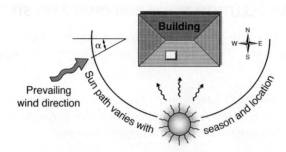

FIGURE 5.6   Orient buildings to take advantage of natural site features such as sun and wind.

The building should be oriented to maximize the positive effects of seasonal variations in solar intensity, solar incidence angle, typical level of cloud cover, and storm influences.

- The building should take advantage of as much daylight as possible in order to decrease electrical lighting requirements.
- Existing or constructed features can shade a building from the sun in hot climates to reduce the solar heat gain and associated air-conditioning needs. Use trees, overhangs, and/or window louvers to shade walls and openings. Light-colored exterior surfaces can help reflect solar radiation, but may cause additional light pollution.
- Natural ventilation will decrease the building's heating and cooling loads. The building can be oriented to align with the prevailing wind direction and to have operable windows so that occupants can take advantage of the natural ventilation. Fans and building openings can also be configured to maximize the natural ventilation available.
- Site features like ponds, fountains, and plants can be strategically located in the path of the prevailing winds to cool the air and provide moisture in hot, dry climates.

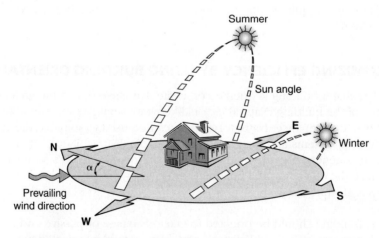

FIGURE 5.7   Design buildings to take advantage of natural solar heating and environmental cooling. Consider the building's heat gain during different seasons and times of day.

FIGURE 5.8 Design overhangs to shade windows during the summer if heat gain is not desired.

REFERENCES AND RESOURCES FOR FURTHER STUDY

1. Illuminating Engineering Society of America (IESNA). This group provides specific lighting standards and requirements for most building types. See http://www.iesna.com for more information.
2. Montoya, Mike. *LEED in Practice—Applying the USGBC Green Building System.* Informa Center for Professional Development, 2006.
3. Sustainable Land Development International (SLDI). *Sustainable Land Development Today: Balancing the Needs of People, Planet and Profit.* Dubuque, IA, 2007.
4. Sustainable Land Development International (SLDI). *Sustainable Urban Development Today: Balancing the Needs of People, Planet and Profit.* Dubuque, IA, 2007.
5. Urban Environmental Institute and Vulcan. *Resource Guide for Sustainable Development in an Urban Environment,* 2007. A case study from South Lake Union in Seattle, Washington.
6. U.S. Department of Energy. *Building Performance Database.* http://www.eere.energy.gov/buildings/database/.
7. U.S. Department of the Interior and National Park Service. *Guiding Principles of Sustainable Design,* 1994. http://www.nps.gov/dsc/dsgncnstr/gpsd/toc.html.

8. U.S. Environmental Protection Agency. The Brownfields and Land Revitalization Technology Support Center. http://www.brownfieldstsc.org/index.cfm.

9. U.S. Environmental Protection Agency Heat Island Effect Initiative. Research results and po-tential strategies from this division of the EPA may be found at http://www.epa.gov/hiri/.

10. U.S. Green Building Council. *LEED for New Commercial Construction and Major Renovations,* Version 2.2 Reference Guide. Washington, DC, 2007.

CHAPTER 6

Managing Site Water Runoff

OUTLINE

- Managing Erosion and Controlling Sedimentation
- Effective Strategies for Stormwater Management
- References and Resources for Further Study

MANAGING EROSION AND CONTROLLING SEDIMENTATION

During the initial development of a site, the removal of existing plants and soil can increase erosion and harm the environment. Erosion occurs when surface soil is loosened, washed away, and redeposited in another location. It is more likely to occur after the site has been graded (i.e., after plants and topsoil that naturally hold the ground surface together are removed).

Erosion can be caused by natural events like rain runoff or wind. Water runoff can wash away unprotected topsoil and expose less stable layers of soil below. Without the protection of a topsoil layer, water runoff is more likely to result in erosion and soil displacement. In addition, wind and runoff can carry the soil to areas where it is not desirable, such as into streams or storm drain systems. The displaced soil can damage the water quality of rivers, streams, lakes, and oceans. Controlling the quality and quantity of water that is released from a site is critical to achieving a good environmental performance. Sediment that escapes a construction site can result in detrimental effects including

- Clogging storm drain systems, potentially resulting in flooding.
- Damaging wildlife habitat and plant life, particularly in streams and lakes.
- Adversely affecting navigation and recreational opportunities in bodies of water like lakes and bays.
- Damaging the quality of water needed to support aquatic organisms and meet drinking water needs.

Plants help control erosion by slowing down the flow of water on the ground surface and by physically holding the soil together with their root structure. Once a site is cleared of its natural vegetation, it is much more susceptible to erosion, which is a considerable concern

FIGURE 6.1 Initial site development increases the potential for erosion and sedimentation.

FIGURE 6.2 Contaminants in rainwater runoff drain into rivers, lakes, bays, and oceans.

FIGURE 6.3 Plants hold soil together. Once they are removed, erosion problems are more likely.

for all construction projects. Erosion also can cause costly rework when any resulting on-site damage has to be repaired.

Permanent erosion control measures like landscaping, swales, and drainage systems are generally included in project designs. However, temporary erosion control measures are often necessary during the construction phase. Temporary systems can manage erosion control and avoid sediment runoff until the permanent measures are in place.

In most jurisdictions where erosion is a concern, local code requires the installation of temporary erosion control measures during construction. For example, in California, any project over one acre must implement a Storm Water Pollution Prevention Plan during construction activities that occur in the rainy season. Erosion and sediment control during construction is a prerequisite to obtain a LEED building certification for projects.

Erosion is generally easy to control with simple technologies. However, more complicated measures may be needed to handle a large grading operation that is conducted when precipitation is expected. Permanent stormwater management systems are generally completed by the civil engineer or the landscape architect during the design phase. Temporary measures are generally the responsibility of the builder. To maximize efficiency and avoid unnecessary duplication, it is critical to involve the builder in the decision-making process of the design phase. Many materials and methods are readily available to control erosion, including the following:

FIGURE 6.4 Vegetated basins can be used for temporary and permanent erosion control.

- Temporary seeding can be used to propagate plants on a slope. The plant roots hold the soil together while the plant mass decreases the water runoff flow rate. Hydroseeding is a method that mixes plant seed with a binding agent that can be sprayed onto the ground. The binding agent protects the soil surface and seeds until the plants are established.
- Wattles are long cylinders made of natural materials such as straw or coconut wrapped in a mesh material. Wattles can be partially embedded in the ground and staked parallel to the graded slope. This orientation will slow down the flow of runoff water and decrease the amount of soil that is displaced.
- Silt fences are made of a geotextile fabric fence generally placed at the toe of a slope, perpendicular to the slope. The silt fence allows runoff water to pass through, but retains the sediment.
- Mulch made of chipped wood, bark, or hay can be placed on the surface of the ground in order to reduce the impact of precipitation as it hits the ground and to slow water flow across the surface of the ground. This effect helps avoid erosion and increases water infiltration into the ground.

FIGURE 6.5   Hydroseeding is used for erosion control on an excavation slope.

• Sediment basins are excavated on-site detention basins into which site water runoff is directed. The basin traps water and allows sediment to settle to the bottom. The sediment can be removed later.

Many additional technologies are available for pollution prevention during construction activity. See the following publications for more information:

• U.S. Environmental Protection Agency *Storm Water Pollution Prevention Plans for Construction Activities.* http://cfpub.epa.gov/npdes/stormwater/swppp.cfm.

FIGURE 6.6   Wattles are used to control erosion temporarily.

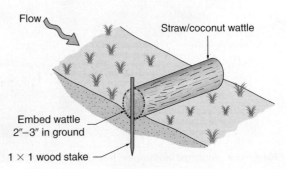

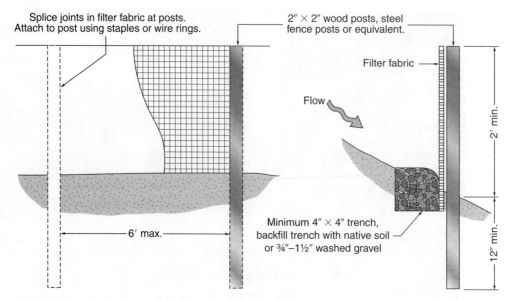

Splice joints in filter fabric at posts.
Attach to post using staples or wire rings.

2″ × 2″ wood posts, steel
fence posts or equivalent.

Filter fabric →

Flow

2′ min.

12″ min.

6′ max.

Minimum 4″ × 4″ trench,
backfill trench with native soil
or ¾″–1½″ washed gravel

FIGURE 6.7   A silt fence is used to control erosion temporarily.

FIGURE 6.8   Mulched landscaping increases runoff infiltration into the ground and aids in erosion control.

- U.S. Environmental Protection Agency Office of Water, http://www.epa.gov/OW.
- State of California *Storm Water Quality Handbooks,* http://www.dot.ca.gov/hq/construc/stormwater/SWPPP_Prep_Manual_3_03.pdf.

EFFECTIVE STRATEGIES FOR STORMWATER MANAGEMENT

As discussed previously, stormwater runoff can have detrimental impacts on the environment. In order to mitigate the environmental impact on streams, lakes, and other water bodies, the amount of runoff that is allowed to leave the site should be minimized and an acceptable level of quality should be maintained in the water that does escape the site. In order to minimize the amount of water that leaves the site, the project design should include site and landscape features that trap the water on-site and allow it to infiltrate into the ground. Generally, materials such as asphalt, concrete, and roofs are impervious and will not allow water to penetrate. To the extent possible, impervious areas should be minimized in a site design. Impervious surfaces may also contribute to pollution in runoff water because they collect oil and gas from automobiles and the oil and gas wash off in the runoff.

Pollutants in runoff water should be treated before the water is allowed to escape the site. The amount of pollution present in water is generally referred to as total suspended

FIGURE 6.9 Retention basins on-site can help control runoff.

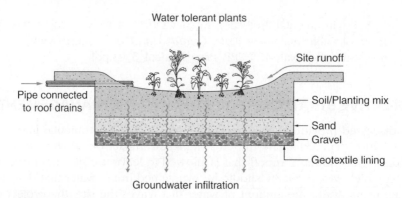

Water tolerant plants

Site runoff

Pipe connected
to roof drains

Soil/Planting mix

Sand
Gravel
Geotextile lining

Groundwater infiltration

FIGURE 6.10 Plants and soil in a well-designed retention basin filter runoff and recharge groundwater.

solids (TSS). Best management practices (BMPs) should be implemented to minimize the amount of runoff that leaves the site and to maximize the quality of any water that does leave. The following are examples of BMPs:

- Rainwater runoff can be channeled into retention basins that are excavated on-site and that become part of the permanent landscape. This practice allows water retained in the basin to infiltrate into the ground.

FIGURE 6.11 Vegetated swales can reduce erosion and sedimentation.

FIGURE 6.12   Porous pavers can serve as walkways and other useful site features.

- Swales are channels that are constructed out of soil and used to direct the flow of runoff water. Swales can be planted with vegetation like grass that will slow the water flow, allow more of the water to infiltrate into the ground, and provide natural filtration of pollutants.
- Porous pavers allow water to penetrate to the ground below. Porous pavers are generally made of concrete and can be used to construct walkways, driveways, patios, and other site features.
- Grass pavers are made of concrete or plastic with open cells. The pavers can support vehicles while plants grow up through the open cells and encourage water infiltration.
- Vegetative roofs are made of plants in a growing medium installed over a waterproof membrane on a building's roof. Rainwater that falls on the roof is retained by the soil and plants.
- Rainwater may be recycled by site and roof drain inlets that direct the water to a storage facility. Processed water can be used for landscaping irrigation and for some uses within the building.

FIGURE 6.13 This typical green roof absorbs rainfall.

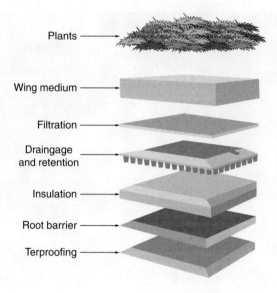

Plants

Wing medium

Filtration

Draingage and retention

Insulation

Root barrier

Terproofing

REFERENCES AND RESOURCES FOR FURTHER STUDY

1. Illuminating Engineering Society of America (IESNA). This organization provides specific lighting standards and requirements for most building types. See http://www.iesna.com.
2. *Small Business Liability Relief and Brownfield Revitalization Act* (Federal Public Law 107–118; H.R. 2869).
3. State of California *Storm Water Quality Handbook.* http://www.dot.ca.gov/hq/construc/stormwater/SWPPP_Prep_Manual_3_03.pdf.
4. Sustainable Land Development International (SLDI). *Sustainable Land Development Today: Balancing the Needs of People, Planet and Profit.* Dubuque, IA, 2007.
5. Sustainable Land Development International (SLDI). *Sustainable Urban Development Today: Balancing the Needs of People, Planet and Profit.* Dubuque, IA, 2007.
6. U.S. Environmental Protection Agency Office of Water, http://www.epa.gov/OW/
7. U.S. Environmental Protection Agency. *Road Map to Understanding Innovative Technology Options for Brownfield Investigation and Cleanup.* http://www.clu-in.org/download/misc/roadmap4.pdf.
8. U.S. Environmental Protection Agency. *Storm Water Pollution Prevention Plans for Construction Activities.* http://cfpub.epa.gov/npdes/stormwater/swppp.cfm.

Improving a Project's Water Use Efficiency

OUTLINE

IMPROVING WATER USE EFFICIENCY

Water-Efficient Landscaping

Clean water is a natural resource, one that is critical to humans, plants, and animals and that is being rapidly depleted due to pollution and human population growth. A building's landscaping can use a significant amount of water, but there are strategies to reduce the amount of water it needs. For example, conducting a soil and climate analysis in the planning stages of a project can help select plant species that will reduce or eliminate the need for irrigation. If water is required to maintain plants, the landscape architect may choose to use only high-efficiency irrigation that can be controlled during seasonal climate changes. This is a schematic design phase activity generally performed by the landscape architect and implemented by the builder.

The following strategies can reduce a landscape's potable water use:

- Choose plant species that are tolerant to the specific climate or microclimates that occur on the development site. Many native or adaptive plants can survive on rainwater alone once they are established. Temporary irrigation can be installed to encourage new plant growth and then removed once the plants are established.

- Turf grasses generally require a significant amount of water and should be used only in areas where they have a functional benefit, such as recreational areas.
- Efficient and effective watering strategies should provide only enough water to keep the landscaping healthy. Water-efficient irrigation techniques such as drip irrigation and micro sprayers direct water only to the specific plant areas that need it. Irrigation systems should be continuously monitored to make sure they are operating properly.
- Irrigation controllers that allow for seasonal adjustments are the best choice. It may be possible to reduce irrigation water needs during seasons with heavy rain.
- Mulch and/or compost can be installed around plants in order to conserve moisture in the soil and avoid surface water evaporation.
- The site contours should be designed to direct rainwater runoff to site planting areas in order to reuse rainwater.

Water-Efficient Buildings

Well-designed buildings use water efficiently to decrease waste. Efficient water delivery systems incorporate strategies to reduce the amount of water used while maintaining a high standard of performance. For example, a low-flow faucet uses less water than a traditional faucet while maintaining its function (providing sufficient water pressure to allow hand washing.) The design of water systems is generally completed by the design team during the schematic design phase of the project.

Following are some of the strategies and commercially available products that can be employed to reduce a building's water use:

FIGURE 7.1 With proper maintenance, waterless urinals use no water and perform as well as conventional urinals.

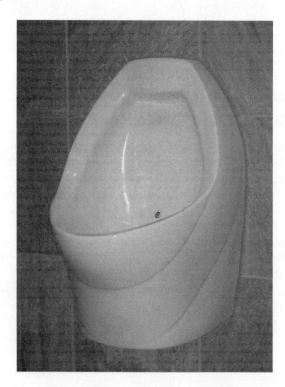

- Low-flow plumbing fixtures use less water for wash basins, toilets, faucets, and showers.
- Commercially available dual-flush toilets allow the user to select a smaller flush (that uses less water) for liquid waste or a larger flush for solid waste.
- Waterless urinals use a chemical in the bowl that is "lighter" than urine and that "pushes" the liquid waste into the drain without using water. A pervasive misconception is that waterless urinals smell bad. It has been demonstrated that smells associated with waterless urinals are usually the fault of improper maintenance. If water is introduced into the bowl in the course of cleaning, it may wash the chemicals down the drain. If this occurs, the system can't function properly.
- Composting toilets convert human waste into an organic compost and usable soil by facilitating the natural breakdown of organic matter into its essential minerals.

The Southface Eco Office (http://www.southface.org) in Atlanta, Georgia, is an example of a notable project that uses multiple innovative strategies for reducing and recycling a building's wastewater.

RECYCLING WASTEWATER

Wastewater from rain runoff and building uses can be reused for purposes on the site and even in the building. The design team and builder must collaborate early in the process and consider site and building uses for wastewater holistically to maximize the chances of success. Wastewater generated from a building is generally classified as rainwater runoff, greywater, or blackwater.

- Rainwater runoff is generated by natural precipitation falling on the site and can be captured with site features and roof drains for reuse.

FIGURE 7.2 Greywater and blackwater have different sources within a building.

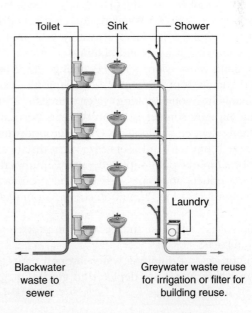

- Greywater is waste from showers and other plumbing fixtures that do not contain feces or food waste.
- Blackwater comes primarily from toilets and sinks and is likely to contain wastes like nitrogen and human pathogens that are not desirable. This type of wastewater needs significant processing before it is suitable for reuse.

In most jurisdictions, rainwater runoff and greywater can be reused for irrigation and within buildings to flush toilets with minimal processing. After more significant processing, blackwater can also be turned into potable water for the building's uses.

A significant amount of water that can be reused for multiple purposes escapes sites and buildings through sewers and storm drains. The demand for water and the costs of processing it are increasing as the world's human population grows and as drinking water sources worldwide become more and more polluted. Reusing wastewater from a building decreases the strain on municipal water processing and is a more efficient use of a quickly diminishing natural resource. Plans to reuse wastewater are generally completed by the civil and mechanical engineers and the landscape architect during the schematic design phase of the project.

Wastewater Retention and Reuse

Natural precipitation such as rain and snow typically falls on a site and is wasted when it leaves the site without being used. Rainwater runoff is conveyed through storm drain systems, along with contaminants such as oil and gas from road surfaces and parking lots, and eventually finds its way into our streams, rivers, lakes, groundwater, and oceans. This pollution affects plant and animal life, disrupts ecosystems, and reduces food sources for humans. For example, many fish species have been exposed to such high levels of pollutants like mercury that they are not safe for human consumption. A notable example of the impacts of rainwater runoff pollution is the Chesapeake Bay, where pollution created a complete dead zone for almost all aquatic life. (The Bay's ecosystem is rebounding now due the implementation of better practices.) Water pollution affects drinking water supplies, food production through agricultural irrigation and watering of animal stock, and recreational opportunities such as swimming, fishing, and boating.

Drinking water sources are becoming scarce in many areas of the world as a result of increased human population and the pollution of existing sources. The process of removing contaminants from our drinking water sources is costly and uses a significant amount of energy. Finding new sources of clean water is becoming increasingly difficult in many areas, necessitating deeper and more extensive groundwater extraction efforts and the pumping of water through pipes to end users over great distances. For example, in many areas of southern California municipally supplied water is pumped from aquifers thousands of miles away through aqueducts and pipe systems. The process of producing and delivering drinkable water to distant locations contributes to environmental damage in several ways:

- The water extraction and distribution process uses energy and has other environmental impacts.
- The process of pumping water through pipes to the end user requires energy.
- Natural resources are depleted to get the materials used in pipelines and processing facilities.

- Ecosystems are affected by the installation of pipelines, particularly in previously undeveloped areas.

Wastewater catchment systems allow the capture and reuse of water from building uses, site runoff, and roof drains for activities such as site irrigation. These efforts decrease or eliminate the need to pipe potable water to the site.

Rainwater Reuse

Rainwater can provide natural irrigation for on-site plants. Excess water can be reintroduced into the groundwater aquifer or stored and reused for irrigation in dry seasons. Rainwater that falls on impervious surfaces like roofs and concrete site features can be captured and reused for landscaping applications. Water from these sources can also be used in the building for toilets and other nonpotable water needs that require little processing. Rainwater catchment systems can be utilized to direct rainwater from the building and site into storage containers. The rainwater then can be used for site irrigation or building uses. Rainwater reuse for irrigation is considered a "first use," and the water generally does not require additional filtration or processing prior to use.

FIGURE 7.3 A direct roof drain to a landscaped area reduces the need to irrigate and recycles wastewater.

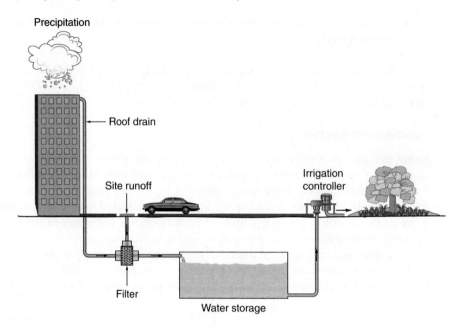

FIGURE 7.4 Wastewater is captured for irrigation.

INFILTRATION SYSTEMS. An infiltration system is designed to capture rainwater and allow it to percolate into the ground to the groundwater aquifer. Generally, storm events produce more water than can soak into the ground. The excess water can be temporarily stored on-site in order to give it more time to infiltrate the ground surface (or in order to reuse it.)

Site water runoff is directed through surface swales or underground pipe infrastructure to a storage area on-site. The water can be stored aboveground in basins or underground in leach tanks. Surface retention basins and swales are integrated into the site and landscaping design and can be installed during construction as temporary measures. Water in these basins and swales naturally infiltrates the ground surface to the underground water aquifers. As the water passes through the ground, surface vegetation and soil naturally remove contaminants from the water and reduce erosion. Basins are sized in accordance with the expected volume of water to be stored and can take up a significant area on a site. Large, deep basins that can store large quantities of water are often not feasible due to site constraints. Water can be stored in underground systems that eliminate or supplement surface basins. Underground storage systems for infiltration generally consist of perforated pipes or tanks with gravel backfill that allow stored water to escape and soak into the ground. Underground systems can be as simple as a traditional leach field or as complex as many commercially available systems.

REUSE SYSTEMS. Stored rainwater can replace potable water for purposes such as landscape irrigation and toilet flushing (both of which require the water to undergo additional filtration). For irrigation, rainwater can be stored and then used during dry seasons. The rainwater is stored in basins or tanks (above or below the ground) and then filtered and pumped to supply the site or building with water. The storage facilities for a reuse system

FIGURE 7.5   Grass-lined swales help rainwater recharge the groundwater aquifer.

are similar to those of infiltration systems, except that they are enclosed and do not allow water to escape. Surface retention basins used for reuse storage can be constructed using natural cohesive soils (such as clay) that are relatively impermeable, but that do lose some water to evaporation. Underground storage systems are common and commercially available.

Greywater Reuse from Buildings

It is often possible to use recycled greywater that is produced by the building for site irrigation and building uses such as flushing toilets. Greywater must be processed before reuse to meet the standards of the federal Department of Health and Human Services or of individual states that have more stringent requirements. While processed greywater is not considered acceptable for human consumption, incidental consumption or exposure is not harmful. Plumbing can separate greywater and direct it to a treatment system. Commercial greywater treatment systems are available that use natural treatment such as live plants, microorganisms and bacteria, or mechanical filtration to clean the water. Once the water has reached an acceptable level of quality, it can be reused for site irrigation and potentially for building uses. Most jurisdictions in the United States allow reuse of processed greywater for building uses such as flushing toilets.

FIGURE 7.6 Greywater may be filtered and reused for building uses such as flushing toilets.

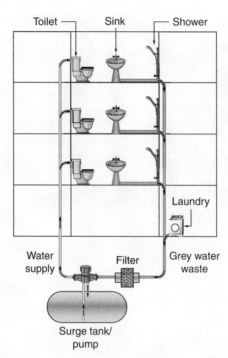

In many areas, municipally treated wastewater is available for site and building use. Wastewater from municipal sewage treatment facilities is often treated to an acceptable level for irrigation and non-potable building uses. It is then distributed to buildings using infrastructure similar to that for potable water distribution. Municipally treated wastewater is generally conveyed through purple pipes to distinguish it from the potable water supply.

REFERENCES AND RESOURCES FOR FURTHER STUDY

1. Southface Eco Office in Atlanta, Georgia. This is an example of a notable project using multiple innovative strategies to reduce the building's water use. http://www.southface.org.
2. U.S. Department of Energy. High Performance Buildings Database. http://www.eere.energy.gov/buildings/database/.
3. U.S. Green Building Council. *LEED-NC for New Construction Reference Guide*, Version 2.2, 1st ed. Washington, DC, 2007.

CHAPTER 8

Improving a Building's Energy Efficiency

OUTLINE

- Efficient Energy Use for Building Systems
  - Heating and Cooling Systems
- Passive Solar Heating and Cooling
  - Direct Heat Gain
  - Indirect Heat Gain
  - Isolated Heat Gain
- Maximizing Energy Performance
- References and Resources for Further Study

EFFICIENT ENERGY USE FOR BUILDING SYSTEMS

Heating and Cooling Systems

Heating and cooling a building can produce pollution and consume significant amounts of energy. Highly efficient systems that use natural features such as the orientation of a building on a site can greatly reduce the building's environmental impact. This activity generally occurs during the design development stage of the project and is a collaborative effort between the mechanical engineer and the architect. Here are some effective heating and cooling recommendations:

- Use only highly efficient building systems. While this strategy may result in higher initial costs, it will likely have long-term cost benefits.
- Use natural features such as building orientation and breezes to reduce the heating and cooling needs of the building.
- Install controls that allow the building systems to be set to maximize efficiency and to adjust automatically on the basis of temperature and environmental fluctuations on a seasonal and/or daily basis.
- Once the building systems have been installed, retain a commissioning agent to test the systems and ensure that they are working as efficiently as possible.
- Establish long-term monitoring and maintenance schedules that will ensure that the system continues to operate efficiently.

- Consider using natural heat produced from geothermal sources. There are many commercially available geothermal heat pumps that draw heat through subterranean ducts to take advantage of the warmer temperatures that generally occur below the ground. (Visit the website of the Geothermal Heat Pump Consortium at http://www.geoexchange.org/.) According to the U.S. EPA, geothermal space conditioning systems represent the most energy-efficient, environmentally clean, and cost-effective methods available (although the report does not include natural methods that are effective, clean, and free).
- Use natural refrigerants such as water, carbon dioxide, ammonia, and propane to cool a building.
- Do not use refrigerants that contain chlorofluorocarbons (CFCs), because they contribute to global ozone layer depletion. Nearly 200 countries, including the United States, have agreed to follow the Montreal Protocol on Substances that Deplete the Ozone Layer and have banned most production and use of CFCs. (Visit the website http://ozone.unep.org/Ratification_status/.)
- Hydrochlorofluorocarbons (HCFCs) cause less environmental damage than CFCs, but they still contribute to ozone depletion and should be avoided if possible. The Montreal Protocol calls for the elimination of HCFC production by the year 2030.
- Hydrofluorocarbons (HFCs) generally result in little or no ozone depletion.

PASSIVE SOLAR HEATING AND COOLING

Passive heating and cooling strategies result in no environmental impacts. In order to incorporate passive techniques into a building design, the design team must consider strategies early in the project. Passive solar strategies can decrease a building's reliance on mechanical space conditioning, reducing the building's energy consumption. Passive design considers the natural process occurring when sunlight shines on building materials. Solar energy radiated by the sun can be absorbed by building materials that store and radiate heat into the interior building space. (If cooling is desired, the building can be designed to reflect the solar energy.)

The amount of heat that a material can store relative to its volume is referred to as volumetric heat capacity or thermal mass. Water has one of the highest volumetric heat capacities and can be used as thermal mass in water walls or roof ponds. Many common building materials such as concrete have high volumetric heat capacities.

Passive solar design strategies generally employ the basic principles of direct, indirect, and isolated heat gain.

Building materials with high volumetric heat capacity:

Water	Rammed earth
Concrete	Sandstone
Brick	Earth wall
Compressed earth blocks	

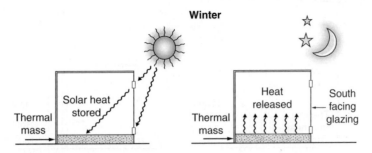

FIGURE 8.1 Solar angle during winter when interior space heat is desired during the day.

Direct Heat Gain

Direct heat gain strategies capture solar heat directly through exterior glazing. Heat energy is absorbed by interior surfaces such as floors and walls made of high thermal mass materials like concrete or masonry. During the day, when heating is desired, sunlight is allowed to enter exterior glazing and is absorbed by interior thermal mass. The heat is stored in the thermal mass and is radiated into the interior space at night.

During times of the year when heating is not desired, such as summer, shade strategies may be used to block the sunlight from shining on the solar mass, which allows the thermal mass to capture interior building heat to help with space cooling.

Well-designed buildings consider the effect of direct heat gain through windows by conducting a heat load analysis of the building. The results are integrated with the building's heating and cooling system design to maximize performance and efficiency of the building's energy use. Generally, the thermal mass is more efficient if it has a large area than if it is very thick. For example, a concrete floor absorbs heat when its surface is exposed to sunlight. Maximizing the square footage of concrete exposed to sunlight is a more effective way to capture heat than increasing the thickness of the concrete floor. A general rule of thumb is that thermal mass materials are effective at absorbing heat only up to about 6 inches. To maximize the potential heat energy absorption, a thermal mass should be in direct contact with the sunlight and not covered by other inefficient storage materials such as carpet.

FIGURE 8.2 Direct heat gain is affected by time of day and by season. This figure shows the solar angle during the summer, when interior space heat is not desired during the day.

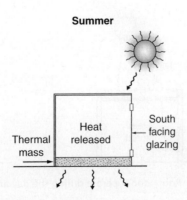

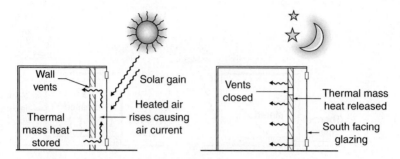

FIGURE 8.3 A Trombe Wall operates during the day and night.

Indirect Heat Gain

Indirect heat gain systems use a thermal mass that is placed between the sun and the interior building space. The indirect thermal mass absorbs heat from the sun, conducts it through the thermal mass, and radiates it into the interior space. Thermal storage walls such as Trombe Walls use exterior glazing with a thermal mass backup wall. Light waves are trapped in the interstitial space between the glass and thermal mass and the heat produced is either absorbed by the thermal mass, radiated by it, or conducted through vents.

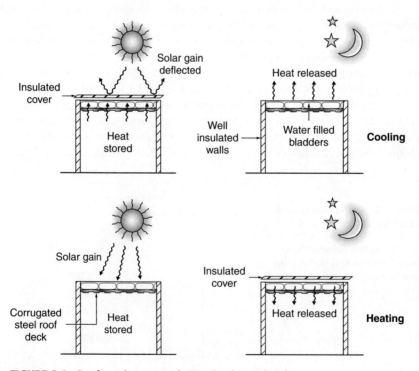

FIGURE 8.4 Roof ponds operate during the day and night.

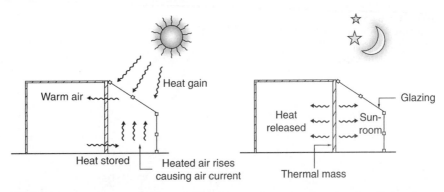

FIGURE 8.5 This sunroom uses an isolated heat gain system that operates during the day and night.

Roof pond systems use water on the roof as a thermal mass. Water is stored in large bladders or containers on the roof exposed to sunlight.

During hot days, the insulation panels are placed between the water mass and sunlight. Later, at night when outdoor temperatures are lower, the panels are opened, allowing heat to be drawn from the building into the thermal mass and to be radiated out of the building for cooling. On cold days, the insulation panels are opened during the day so that the water mass may absorb heat from the sun. The panels are closed at night so that the heat can be radiated into the building to provide heating. The heated water can also be used in combination with a hydronic heating system to provide building conditioning.

Isolated Heat Gain

Isolated gain systems use a heat-collecting space that is separate from the interior space, such as an atrium or solar greenhouse. Isolated gain systems use the principles of direct and indirect gain systems for space conditioning. Generally, sunlight enters the space and is stored in a thermal mass wall and in the air of the atrium or solar room. The thermal mass wall can store the heat and conduct it into the interior building spaces. This process may be improved by installing vents in the thermal mass wall that can draw or expel heated air from the interior space. Vents may use efficient mechanical fans or naturally occurring convection currents. (Heated air rises and produces an air current.)

MAXIMIZING ENERGY PERFORMANCE

The use of energy-efficient lighting, appliances, and equipment can significantly decrease the amount of energy that a building consumes. In a conventional building, energy is consumed from heating and cooling large volumes of air, heating large volumes of water, and distributing electrical power. Ideally, a building's design considers the specific use of each space and then minimizes the energy consumption for the planned application. Architects and engineers must collaborate early in the design process in order to maximize energy performance for a building project. To improve the building's environmental performance, every project should maximize natural methods that limit pollution. Natural and mechanical systems must be designed to work together to maximize the overall efficiency and environment performance

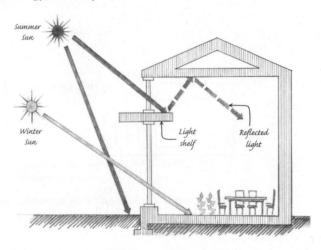

FIGURE 8.6 Skylights can be combined with light shelves and reflective interior wall finishes to maximize light exposure indoors.

of the building. Many strategies are available to maximize the energy efficiency of a building and reduce its heating and cooling costs, including the following:

- Orient the building to take advantage of natural shading and solar heat gain. This feature will minimize the need to use mechanical heating and cooling systems.
- Take advantage of natural light to decrease the energy consumed by artificial lighting. Use design strategies such as atriums, large exterior windows, high ceilings, building elements that redirect daylight and control glare, and skylights.
- Decrease unwanted heat gain through windows by using low-emissivity (low-E) glass.
- Install fluorescent light fixtures instead of incandescent units that consume significantly more energy to produce the same amount of usable light.
- Install point-of-use controls in individual areas that illuminate only the light that is needed for each space.
- Control light and heat gain through windows by using thermochromic glass (glass that turns darker when warmed by the sun), photochromic glass (glass that turns darker when exposed to bright light), or electrochromic glass (glass whose darkness can be controlled by an electronic switch).
- Install only energy-efficient appliances and equipment. The U.S. EPA certifies appliances that meet specific environmental criteria. Look for appliances that have an "Energy Star" rating. (Visit the website http://www.energystar.gov for more information.)
- Install lighting control systems that automatically operate lights based on programming events such as time of day, sunrise, sunset, occupancy, and amount of available sunlight.
- Install wireless lighting control systems that integrate control signals from wireless switches, computers, and occupancy sensors to operate lighting for the building.
- Install "smart" automated interior or exterior shading devices such as blinds and shades designed to respond to changing solar conditions using a network of sensors and computer-controlled shading devices. Shading devices help avoid direct solar heat gain to interior spaces during times of day or seasons when heat is not desired.

REFERENCES AND RESOURCES FOR FURTHER STUDY

1. Southface Eco Office in Atlanta, Georgia. This is an example of a notable project using multiple innovative strategies to reduce the building's water use. http://www.southface.org.

2. U.S. Department of Energy. High Performance Buildings Database. http://www.eere.energy.gov/buildings/database/

3. U.S. Environmental Protection Agency. Energy Star program. http://www.energystar.gov. The EPA certifies appliances that meet specific environmental criteria. Look for appliances that have an "Energy Star" rating.

4. U.S. Environmental Protection Agency. *Space Conditioning: The Next Frontier—Report 430-R-93-004.* 1993.

CHAPTER 9

Using Renewable Energy Sources

OUTLINE

- Renewable Energy Sources
- Green Power
  - Wind Farms
  - Photovoltaic Power Plants
  - Solar Thermal Power Plants
  - Geothermal Power Plants
  - Hydroelectric Power Plants
    - Tidal-Surge–Powered Electricity
  - Biomass Power Sources
  - Tidal Power Sources
- References and Resources for Further Study

RENEWABLE ENERGY SOURCES

Renewable energy systems use technologies designed to capture energy from natural, non-polluting sources like the sun, wind, geothermal heat, and water. Many power plants use renewable energy systems to supply green power to communities. Some specific buildings or sites use individual power generation devices. On-site renewable energy can produce power for an entire building. Power generated this way is often supplemented by energy from the power grid during times of low production or when on-site power storage using batteries is not an option. If on-site generation produces more power than the building can use, excess power can be fed back into the power grid and used by the power company for other buildings on the grid. The use of renewable energy produces little or no pollution and can result in significant energy cost savings over the life of a building. There are many viable commercially available renewable energy systems that can be used for building power generation.

Globally, power is generated by the following sources:

- Fossil fuels 67%
- Nuclear 14%
- Large hydropower 15%
- New renewable energy sources 3.4%

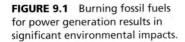

FIGURE 9.1 Burning fossil fuels for power generation results in significant environmental impacts.

The majority of the world's power supply comes from burning fossil fuels, a practice that results in significant pollution, primarily from carbon dioxide emissions, which contribute to global warming.

High oil prices, decreasing fossil fuel availability, and climate change concerns have spurred the development of government initiatives and legislation to encourage the use of more renewable energy sources. For example, the European Union agreed in 2007 that 20% of energy production in EU countries should come from nonpolluting sources by 2020. The $2.8 billion California Solar Initiatives aim to create 3,000 megawatts of new solar power by 2017. The renewable power industry has seen unprecedented growth in the last five years. According to the report by the Renewable Energy Policy Network for the 21st Century, worldwide investment in renewable energy (new renewable energy capacity, manufacturing plants, and research and development) increased from $80 billion in 2005 to $100 billion in 2006. Some of the world's largest power companies, including Royal Dutch Shell, General Electric, and British Petroleum, have made commitments to provide renewable energy.

Nuclear power production effectively causes no greenhouse gas emissions. However, the mining of raw uranium and the storage of hazardous waste are a primary environmental concern. Nuclear power is not represented here as a renewable energy source, due primarily to the long-term storage of dangerous waste that is necessary and to the cost and environmental impacts of uranium mining. Until nuclear power can be produced at a reasonable cost without significant government subsidies, it is not a practical source of power. Until the nuclear power industry *decides* to recycle waste, nuclear energy will not be a green source of power. Such recycling is an option; for example, France recycles used uranium from nuclear power plants, effectively eliminating the need to mine raw uranium and significantly reducing the necessary amount of waste storage.

GREEN POWER

Electrical power generation can be provided by a municipality or power company. This type of off-site power is generated by a power plant and conveyed to buildings through conductors. It is commonly referred to as "grid"-supplied power. The majority of the world's power plants burn fossil fuels that cause significant pollution. However, there are many commercially viable technologies to provide renewable energy, or green power, on a large scale that are successfully being used throughout the world. This type of off-site generated power is referred to as "green power." Renewable energy sources have a high potential to meet power needs around the world. New power sources are necessary to support the world's population growth. As a practical source to support the world's future energy needs, renewable energy exceeds all other readily available sources, including the continued use of fossil fuels.

Many current trends are encouraging the competitiveness of green power strategies. Some of the incentives are

- Net metering laws, which give credit for electricity fed into the grid.
- Feed-in tariff laws, which allow building owners to receive revenue for excess power generation. (Feed-in tariff laws in Germany are currently the main driver of the world's photovoltaic growth.)
- Rebates and tax credits at the federal, state, and local levels, to encourage consumers to implement sustainable power strategies.
- Government grants for fundamental research in technology to make production cheaper and improve efficiency.
- Green loan programs with low interest rates that may decrease deployment costs.

Wind Farms

Wind power converts wind energy into electricity by using generators driven by wind turbines. Wind power generators use natural wind currents to turn turbine blades (which most commonly look like an airplane propeller). The blades run a generator to produce electricity. Wind power production produces little or no pollution. The blades are generally mounted on a mast to take advantage of the stronger and more consistent wind patterns that occur high above the ground surface.

Wind farms group multiple turbines in the same location to produce electric power on a large scale and are being used throughout the world. The turbines of a wind farm are connected to medium voltage networks. The electrical current is "stepped up" to high voltage by means of a transformer and is fed into the power grid. Wind farms are generally located in areas with high wind speeds and relatively predictable wind pat-

FIGURE 9.2   A wind farm consists of multiple wind turbines that can provide municipal power to communities.

terns. Higher elevations and ridges in hilly and mountainous areas are generally preferred because these sites can take advantage of higher wind speeds and maximize the efficiency of the power production. The wind turbines at such sites benefit from topographic wind acceleration, an increase in wind speed that occurs as the wind travels over a ridge.

Construction in these areas is generally more expensive, but this initial cost is offset by the greater efficiency of the turbines. Modern turbine design considers and mitigates the impacts on birds and other animals, and these turbines have a small environmental footprint. However, wind farms may require a significant minimum area to be commercially viable and may have impacts on the surrounding community related to noise, land use, impairment of scenic views, and damage to the ecosystem. Wind farms commonly supplement another power source instead of serving as a primary power source, since the variability of wind output and the lack of power storage capacity make their output somewhat unreliable. Wind speeds are difficult to predict, inherently variable, and often seasonal, and they cannot guarantee continuous power. Though promising technology for the storage of wind energy is emerging, it is not yet commercially viable, so power produced by a wind farm must be used immediately, or it will be wasted.

Offshore wind farms use large wind turbines installed in the ocean or large lakes connected to the grid through high-voltage direct current undersea cables. Offshore wind farms take advantage of

- Greater wind speeds (on the order of 90% greater than on land) that occur over open water.
- Calmer surface conditions (compared with near-shore installations).
- The remote location mitigates a large turbine's inherent obtrusiveness and noise.

Offshore wind turbines are generally very large and are more expensive to install and maintain than land-based turbines. The foundations may be in very deep water and are designed to support the turbine against water currents and wind forces. The mast is taller than land-based masts due to the water depth (near-shore turbines don't need quite such a tall mast, but are much less efficient at producing power than turbines that are further offshore). The exposure of the mast to water increases maintenance costs, and corrosion protection is necessary, especially in saltwater environments.

Many countries and regions recognize the potential for offshore wind power. The United Kingdom plans to use offshore turbines to power every house in the country by 2020. In the United States, there are plans to install freshwater wind turbines in the Great Lakes and Pacific Northwest. Denmark has Continental Shelf offshore wind farms. One of the most notable offshore wind farms is the Arklow Offshore Wind Power Plant in Ireland. This wind farm produces 25 megawatts (MW) of power and reportedly provides power for 16,000 homes. The farm uses massive 3.6 MW turbines, currently the largest commercially available turbines, built by GE Energy. Each turbine is over 40 feet tall and is installed on tubular steel foundations driven with a hydraulic hammer. The mast is 16 feet wide at its base and the rotors are over 340 feet in diameter, comparable to the wingspan of a Boeing 747 airplane.

By 2010, the World Wind Energy Association expects 160 GW of capacity to be installed worldwide, up from 74 GW at the end of 2006. This increase implies an anticipated net growth rate of more than 21% per year. World wind generation capacity more than quadrupled between 2000 and 2006. It is currently estimated that, worldwide, wind power has the potential to generate over 74,000 MW.

Photovoltaic Power Plants

A solar photovoltaic (PV) cell is used to convert the sun's light into electricity and is used to power everything from calculators to entire communities.

The sun produces a significant amount of energy that is transmitted to the earth through solar radiation. This solar energy is partly reflected and absorbed by the earth's upper atmosphere. Clouds, smog, dust, and other atmospheric situations increase the amount of energy reflected and absorbed before it reaches the earth's surface. High in the atmosphere, the energy is diffused and filtered into different light spectrums, primarily the visible electromagnetic spectrum and infrared spectrum. These factors affect the amount of light and energy that is available for photovoltaic power generation on the earth's surface. Of this energy that penetrates the atmosphere and its pollutants, approximately half is absorbed by land and water masses. The remaining solar energy is reflected and reradiated into space. The energy that is absorbed supports life at the very basic level through

FIGURE 9.3 Photovoltaic solar panels convert the power of the sun into a usable form.

photosynthesis and is abundant enough to provide all of the earth's electricity needs. Enough sunlight falls on the earth's surface each minute to meet world energy demand for an entire year if the sunlight were all captured and used without waste. The embodied energy necessary to produce common solar cells is significantly lower than that necessary for traditional methods of electricity generation. For example, silicon from one ton of sand can produce as much electricity when it is used in photovoltaic cell production as could burning 500,000 tons of coal.

PV panels are made up of individual cells wired together and mounted on a surface such as glass to form a panel. PV panels are wired together and connected to an inverter or a storage medium such as batteries. Photovoltaic power plants use an array of PV panels that are designed and placed to maximize sun exposure. Electricity produced in this way generally supplements the municipal grid supply.

PV power plants need a significant amount of space to be commercially viable, and there are very few of them. Smaller systems that support small communities or individual buildings are common in most countries around the world, and their prevalence is growing exponentially. Government incentives, decreased cost, and technological advancements have made solar a viable power source with a payback period that is generally within acceptable investment strategies. Many communities around the world are using feed-in tariffs

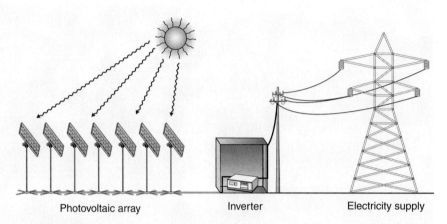

FIGURE 9.4   Photovoltaic power plants can supply the municipal power grid.

to support PV installations on buildings. These tariffs effectively allow the user to sell power back to the grid, making each project a power plant. For example, the city of Freiburg, Germany, is installing PV panels on building roofs to provide power for each unit. Systems are designed to produce more power than the community needs so that they can "sell" the excess power back to the power company. The revenue generated from the excess power is used to offset the initial investment cost and is invested in the long-term maintenance needs of the PV plant. The amount of the world's electricity that is supplied by solar is expected to increase more than 50% (on the order of 3,000 MW) by the end of 2008.

Some of the world's most notable photovoltaic power plants include the Waldpolenz Solar Park in Germany, consisting of over 550,000 solar panels that supply 40 MW of electricity, and the Serpa solar power plant in Portugal, which produces 11 MW using 52,000 PV panels distributed over 150 acres.

Solar Thermal Power Plants

Solar thermal energy is produced by concentrating sunlight using mirrors or lenses into a small beam that is used to produce steam to run electricity generators. Mirrors and lenses are aligned and controlled to track the sun as it moves across the sky. This system is generally useful only in areas that have a great deal of daily sun and plenty of open space, such as desert areas. The environmental footprint created by solar thermal power plants is less than that needed for PV, hydroelectric and coal burning. This concentrated solar light can produce extreme temperatures (on the order of 1,500C), the heat from which is used to run electricity generators. While PV cells can be used to convert the light into electricity, it is more efficient to use a heated fluid to power a generator. Most modern solar thermal power plants use molten salt to collect the concentrated heat and use steam to run electricity generators.

Solar thermal plants concentrate light into a receiver that converts the light to heat or electricity by using a reflecting surface such as a solar trough, linear parabolic reflector, parabolic reflector, or heliostat (flat reflector).

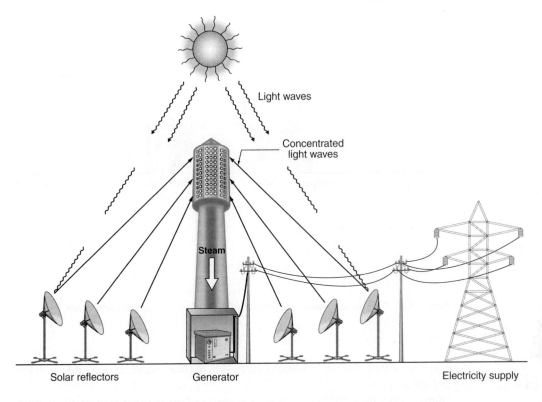

FIGURE 9.5 A solar thermal power plant concentrates solar power by using mirrors or lenses.

The largest solar thermal project in the world is the SEGS power plant facility in the California desert, which uses over 1,000,000 mirrors covering more than 1,600 acres (6.4 km²) and produces more than 350 MW of solar power. Another notable project is the PS10 Solar Power Tower in Seville, Spain, which produces 11 MW of electricity with over 600 heliostats and plans to expand to over 300 MW by 2013. Each reflector is 120 square meters (1,292 square feet), and the receiver is a 35-story tower that runs a steam turbine.

Geothermal Power Plants

Geothermal energy is produced from the heat that naturally occurs below the earth's surface. In some areas, this heat from the earth's core is relatively close to the surface and can be tapped for power generation. Steam from subsurface ground fractures or hot water from subsurface sources is used to power generators and produce electricity. Excess heat and fluids are injected back into the ground source.

Geothermal heat sources can be used for large-scale power production or for smaller site applications. Geothermal power plants operate throughout the world. The Geysers Power Plant in California, the largest geothermal development in the world, consists of over 20 power plants that produce electricity from over 350 deep-ground steam wells and together

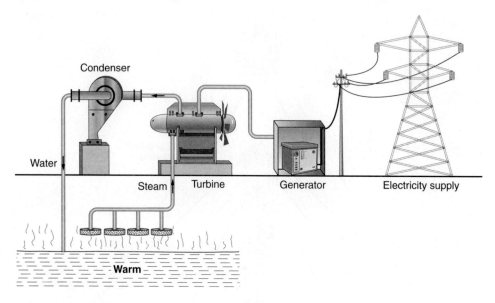

FIGURE 9.6   Geothermal power generation uses heat created by the earth's core.

produce 750 MW of power. In Iceland, over 26% of total electrical demand is generated from geothermal sources.

Geothermal heat produces minimal amounts of air pollution (primarily carbon dioxide, nitric oxide, and sulfur), on the order of 5% of the emissions from fossil fuel power plants. Geothermal heat extraction and the reinjection of waste materials may affect land stability and the heat available may decrease over time.

Hydroelectric Power Plants

Hydroelectric power uses the natural power of water movement to produce electricity. The most common way to generate hydroelectric power is to dam water in rivers and use its potential energy to drive turbines that produce electricity. Hydroelectricity has been used throughout history and now produces on the order of 15% of the world's electricity.

While hydroelectric power generation produces very little pollution, dams can have significant impacts on ecosystems and aquatic life.

An example of hydroelectric power generation is the Three Gorges Dam across the Yangtze River in China, under construction at the time of this writing. This project is intended to produce over 22,000 MW of electricity, which will make it the largest hydroelectric plant in the world.

TIDAL-SURGE–POWERED ELECTRICITY.   Hydroelectric power generation is not limited to dams. New technology in the form of underwater turbines is allowing us to generate power from the flow of rivers and from the ocean's constant wave motion. These underwater turbines are powered by the current produced in a river or by ocean tides. Tidal energy turbines are being explored throughout the world. One facility in New York's East River has the potential to supply enough power for 8,000 homes (on the order of 8 KW); another in the Solomon Islands is a 50 KW plant. While the technology and implementation of tidal-surge–powered electricity is relatively unexplored, it has great potential as a long-term sustainable power source.

Biomass Power Sources

The most basic way to produce power from biomass is to burn wood for heat. Biomass power systems burn biological materials and biodegradable waste to produce energy. These systems do not include organic materials (such as coal) that have been transformed by a geological process. For industrial production, biomass is most commonly produced from plant matter such as corn, poplar, willow, sugarcane, and hemp; it also may be generated from post-consumer biodegradable waste, such as cooking oil, that can be burnt as fuel. Biomass is generally considered to be a sustainable power source because it is more efficient than most traditional methods and relies on raw materials that can be grown.

Biological materials also are used to produce fuel to replace petroleum-based fuel for vehicles. Biofuel, or agrofuel, produces fewer emissions from vehicles.

However, biomass power still creates emissions that contribute to air pollution and global warming. Similar amounts of CO_2 emissions are produced from biomass (primarily from the burning of fossil fuels in order to grow and fertilize the crops and process them into fuel) as are produced from fossil fuels. Some environmentalists question the use of plant matter for fuel because its production may have negative environmental implications. For example, because land to grow biomass could have been used to grow food crops, biomass production can contribute to high food prices. It also is associated with soil erosion, deforestation, and human rights abuses.

Tidal Power Sources

Tidal power uses the water movement from ocean tides and the rise and fall of the ocean surface to power electricity generators. One tidal power generator is Oregon State University's SeaBeav project, which uses ocean-deployed buoys attached to the ocean floor miles from shore. These buoys include a magnetic shaft and electric coils that produce electricity as the buoy rises and falls with the surface of the ocean. Microhydroelectric turbines can be placed in tidal waters to take advantage of the constant movement of water due to tidal action. The study of tidal power from ocean currents and movements is in its infancy, and this system is not widely used, but it has great potential for future sustainable power generation.

REFERENCES AND RESOURCES FOR FURTHER STUDY

1. Davidson, Sarah. National Survey Report of PV Power Applications in the United Kingdom, 2006.
2. Dunlop, Jim. *Photovoltaic Systems.* Homewood, IL: American Technical Publishers Inc., 2007.
3. Geothermal Heat Pump Consortium. http://www.geoexchange.org/.
4. Renewable Energy Policy Network for the 21st Century (REN21). *Renewables 2007 Global Status Report,* 2007. http://www.ren21.net/globalstatus-report/default.asp.
5. Southface Eco Office in Atlanta, Georgia. This is an example of a notable project using multiple

innovative strategies to reduce the building's water use. http://www.southface.org.

6. U.S. Department of Energy. High Performance Buildings Database. http://www.eere.energy.gov/buildings/database/.

7. U.S. Environmental Protection Agency. Energy Star program. http://www.energystar.gov. The EPA certifies appliances that meet specific environmental criteria. Look for appliances that have an "Energy Star" rating.

8. U.S. Environmental Protection Agency. *Space Conditioning: The Next Frontier—Report 430-R-93-004.* 1993.

9. Wissing, Lothar, Jülich, Forschungszentrum & Jülich, Projektträger. *National Survey Report of PV Power Applications in Germany 2006, Version 2,* 2007.

10. Zhai, X.Q., R.Z. Wang, J.Y. Wu, J.Y. Dai, and Q. Ma. Applied Energy, Design and Performance of a Solar-Powered Air-Conditioning System in a Green Building. *Renewable Energy Resources and a Greener Future,* Vol. VIII-7-1. Shenzhen, China: ICEBO, 2007.

CHAPTER 10

On-Site Power Generation Using Renewable Energy Sources

OUTLINE

PHOTOVOLTAIC (PV) POWER GENERATION

Photovoltaic (PV) cells convert sunlight into electricity. PV cells are wired together and installed on panels often generically referred to as solar panels. PV panels collect sunlight and convert it to electric power that can supply electricity for a building project. The basic components of on-site PV systems include PV cells, panels, batteries, and conductors.

PV cells are most commonly made from silicon (single crystal, polycrystalline, or amorphous). However, some types of PV cells are manufactured from cadmium. Silicone and cadmium are photodiodes and can create electricity that is produced entirely by the sun's transduced light energy. Photons in sunlight cause the electrons in the photodiode to move faster, which produces an electric current. This current is transmitted by conductors in the form of direct current (DC) which can be used to recharge batteries or power equipment. Solar-supplied DC electricity is commonly used to power individual systems such as pumps

FIGURE 10.1   A photovoltaic panel is mounted on a building's roof.

or lights directly. To provide power for building use, the DC current is converted to an alternating current (AC) by an inverter.

A single PV cell can be used to provide electricity for small items that do not require a significant amount of energy, such as on-site emergency phones and lights. To provide enough power to support a building, single PV cells are electronically connected to form an individual module called a solar panel. To protect the PV cells, panels generally sandwich multiple cells between two plates of glass. Aluminum frames or braces are used to further protect the panes and provide a means for supporting and connecting them to the structure. Multiple solar panels forming an array are used to provide electric power to meet a portion or all of a building's electric energy needs.

The number of panels needed for a specific project depends on the size and energy consumption of the building and the efficiency of the system selected. Many publications and design aids are available for evaluating solar electric power generation and for designing a PV system. The design process is highly dependent on the efficiency of the panels selected and is generally completed by the PV panel system manufacturer or supplier.

PV electricity is a point-of-use power source, meaning that any power generated that is not immediately used is wasted. PV panels lose efficiency when sunlight is obscured, such as on cloudy days, and they are ineffective when they are not exposed to sunlight, such as at night. Batteries and grid-connected systems can be used to store or use the excess power generated and can provide electricity when the PV panels are not operating.

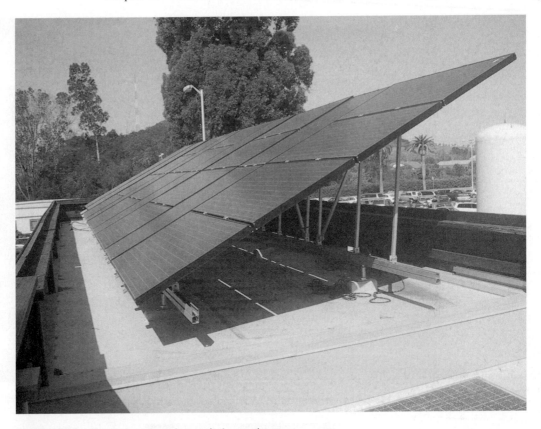

FIGURE 10.2 Shown here is a photovoltaic panel system.

Electricity Storage with PV Systems

PV systems can be connected to batteries that can store excess energy produced during sunny times for use at night and on cloudy days. The batteries used in most PV systems are commonly available. Battery storage is often not an option for larger building projects because a lot of space is needed to store enough power to meet the needs of an entire building. Battery storage strategies are generally used in buildings in remote locations that are without access to the public electricity supply grid.

Grid-Connected PV Systems

On-site PV panels can be wired to an inverter that converts the electricity so that it can be connected to the municipal electricity supply grid. Grid-connected on-site solar electricity systems allow excess energy produced during the day to be fed back into the grid to supplement its power. Conversely, at night or on cloudy days, if the PV panels' production decreases, it can be supplemented by grid power.

On-site electricity can be generated using PV panels, which are most commonly installed on the building's roof and/or exterior walls. Panels are oriented towards the prevailing sun direction and angle depending on the location, in order to maximize the amount of sun exposure. Because PV panel energy production drops on cloudy days and at night, PV panels may not provide a constant amount of electricity for a building's use.

FIGURE 10.3   Batteries are used to store photovoltaic-produced electricity.

FIGURE 10.4   This photovoltaic system has an inverter for a grid connection for a small project.

Building-Mounted PV Systems

Building-mounted PV arrays are currently the most widely used installation method. These arrays are generally mounted on the roof or exterior wall surfaces of buildings. Building-mounted PV panels have mechanical mounting systems that are physically attached to the exterior building skin or roof, which makes them a natural choice for remodeling work.

The number of panels in a building-mounted array is limited by the surface area available on the building. The building surfaces for mounting are limited to areas where the panels do not conflict with the operation of the building exterior. Space on the side of buildings is limited to areas without features such as windows; panels mounted here will affect the architectural aesthetics of the building. Wall-mounted PV systems are often installed on shear walls and on elevator and stairwell walls that do not have window openings.

Building-mounted panels are configured in a direction that takes advantage of the sun location, angles, and intensity on different sides of the building and during different seasons. Roof spaces can be used to mount PV panels, except where there are obstructions such as skylights or HVAC equipment. On large commercial projects, there is generally not enough square footage of roof area to provide the power needed to support an entire building. Often the PV system must be supplemented by electricity supplied from the public grid. Building-mounted PV panels are generally mounted directly on the exterior skin of the building. Since the building's skin is its defense against the elements, special attention is required during design and

FIGURE 10.5 These photovoltaic panels also serve as an awning.

installation in order to avoid causing problems such as roof leaks and air intrusion. For retrofit work, the building structure must be analyzed to evaluate the effect of adding the weight of the panels. Building-mounted PV panels are commercially available from multiple suppliers who can assist with the connection requirements and design.

Site-Mounted PV Arrays

Panels similar to those used for building-mounted arrays can be incorporated into site features and connected to the building systems through underground transmission lines. PV panels can be mounted on architectural site and building features such as fences, parking covers, overhangs, and awnings. They also can be used to provide electricity for stand-alone devises such as parking meters, emergency phones, and building and traffic signage.

Some innovative designers are including solar panels on artwork such as solar "trees," which have a treelike structure made of steel, and "leaves" made from solar panels. Site-

FIGURE 10.6 Solar PV panels can be used to provide remote power for site features such as signs.

FIGURE 10.7   Shown here is a solar-powered station for recharging and storing electric bicycles.

mounted PV arrays can provide power for a structure via underground transmission lines or can power a stand-alone device separate from the structure, eliminating the need to provide underground electricity from the building.

Building-Integrated Photovoltaics

Building-integrated photovoltaics (BIPV) are used in place of, or integrated with, common building components such as roofs, facades, and glazing. The following are examples of BIPVs:

- Thin-film cells are made in flexible sheets (rolls) similar to traditional roof membranes. The silicone used to make thin-film cells in current technology is "amorphous" and needs less light to function effectively, allowing less silicon to be used in each cell compared with the amount needed in traditional PV panels. As a result, the film is lightweight with a thin profile. Sheets of thin-film cells can be integrated with the traditional roof membrane on low-slope roofs or adhered to the channels of standing seam roofs. They are commercially available from companies such as BP Solar and Open Energy.
- Solar shingles are individual PV modules that are shaped and designed to mimic traditional roofing materials such as tile and asphalt shingles. Solar shingles are made of

high-efficiency PV cells that are integrated with a multilayered laminate material that may contain glass, polymers, or plastics. Solar shingles are the most widely used BIPV and are available from companies such as Ok Solar and UNI-SOLAR.

- Solar building facades are panelized PV systems that can be used for a building's exterior façade. PV cells are integrated into steel and composite sheets with high-density foam insulation that can be used as panels for exterior wall surfaces such as curtain wall systems. This technology is new and quickly evolving as demand increases; currently, it is being used primarily in Europe. Some well-known examples of BIPV solar facades include Freiburg Main Train Station, ThyssenKrupp Stahl AG in Duisburg (5,000-square-foot photovoltaic façade), Tübingen sports arena, and BioHaus office building in Paderborn, all in Germany; and the CIS Tower in Manchester, England.

- Solar glazing involves thin-film solar cells sandwiched between plates of glass. This glazing panel can be integrated into the building envelope in place of glass glazing. The currently available solar glazing panels are semi-transparent and generally are not practical for view windows. Solar glazing panels can be used for other building elements as well, such as skylights, cladding, awnings, spandrel glass, and curtain walls. There are tremendous growth opportunities for solar glazing, and some products have already enjoyed great success, including PowerGlaz crystalline silicon cells by Romag, ASI® glass solar modules by Schott, and amorphous silicon solar glass panels by XsunX, Inc.

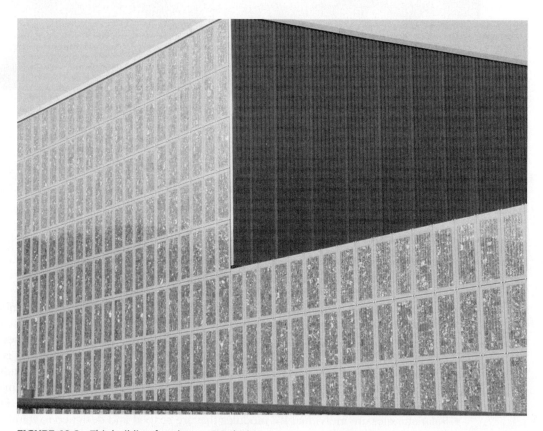

FIGURE 10.8   This building facade uses PV glazing.

Other companies around the world that are researching and developing solar glazing products include SunPower Corporation, Atlantis Energy, EVOENERGY, and Stellaris.

While BIPV systems are commercially available and are being used on buildings around the world, the technology is in its infancy and is expensive relative to other types of PV systems and traditional fossil fuel power sources. This initial high cost of BIPV modules is offset by cost savings over time. The long-term savings from the reduction or elimination of electricity costs for the life of the system offsets the initial higher purchase and installation cost of BIPV systems. The return on investment (the amount of time it takes for the long-term cost savings to pay off the initial investment) for BIPV systems is generally longer than is practical for most building investors. However, the solar cell industry is realizing its highest growth rate in history and is becoming more affordable as the technology catches up with the demand. On the basis of projected continuous increases in energy costs, increased consumer demand, and government subsidies and incentives, the outlook for BIPV systems is positive. Several countries, including France, Italy, Germany, and the United States, have monetary incentives that are intended specifically for BIPV systems (as opposed to other PV systems). Programs in the United States vary by state; more details about them are available from individual state energy offices or the National Association of State Energy Officials (http://www.NASEO.org).

Financial Incentives for On-Site Solar Power Generation

Public policies on solar energy are being implemented around the world to decrease emissions, encourage industry growth, achieve energy independence, and stimulate high-tech job growth. The goal of public incentives regarding solar energy is to achieve grid parity, the point at which PV electricity becomes less expensive than grid-supplied electricity. Reaching grid parity generally requires sufficient amounts of sunlight coupled with a high cost for grid electricity. Italy and Hawaii are two areas that have achieved grid parity. Many countries have initiatives in place and are working toward the goal of achieving grid parity. The United States has a plan to achieve grid parity by 2015. (Italy achieved it in 2006.) Most public incentives are in the form of subsidies, feed-in tariffs, and net metering.

SUBSIDIES.   Subsidies include a refund from public agencies or from electric companies for installing PV systems. Well-known subsidy programs include the California Solar Initiative and the Ontario Standard Offer Program.

NET METERING.   This incentive is generally offered by the public electricity supplier and allows excess electricity generated by on-site, grid-connected systems to be traded for grid-supplied electricity. This strategy ideally allows buildings to produce excess electricity and feed it into the grid, earning credit that can be used to "pay" for grid-supplied power at night or on cloudy days.

FEED-IN TARIFFS.   Under the feed-in tariffs system, a building sells its excess electricity back to the utility company. Such a building is known as a plus-energy development that produces more power than it can use. These developments can generate revenue from their excess electricity generation—revenue that helps them offset their grid electricity use at night and pay for long-term maintenance. One of the world's most notable plus-energy developments is the Vauban District housing development in Freiburg, Germany. The Erneuerbare Energien Gesetz (EEG) law, introduced in Germany in 2004, was the first large-scale feed-in tariff system. Germany's EEG law resulted in explosive growth in the PV industry, and Germany currently has the highest solar PV capacity in the world. After the unprecedented success of Germany's feed-in tariff law, other countries, such as Spain, Italy, Greece, and France, have followed suit.

BIOMASS POWER SYSTEMS

Biomass generators produce electricity (or heat) by burning a fuel source that is made from processed natural products such as used cooking oil and harvested plant matter. First-generation biomass generators have a reputation for producing amounts of CO_2 emissions (primarily from growing crops and processing them into fuel) similar to those produced by fossil fuels. However, there are many promising products entering the market that reportedly can burn almost anything, even trash, and can produce power that passes emissions standards. The most likely market for these products is buildings that would otherwise use electricity and heat produced by diesel generators. One biomass generator company, California-based AgriPower Incorporated, is currently testing and planning to manufacture a biomass generator that uses a high-temperature sand bed to vaporize biomass materials within a few seconds with reported lower emissions for the system. There are many commercially available agricultural fuel products for biomass generators, such as compressed wood pellets made by Polar, and plenty of locally available sources for post-consumer biodegradable waste, such as waste cooking oil from restaurants.

GEOTHERMAL POWER SYSTEMS

Geothermal energy is produced from the heat stored beneath the Earth's surface using geothermal heat pumps. On-site geothermal heat pumps can run generators to produce electricity or directly heat and cool a building. A thermal transfer medium such as water is pumped through an underground pipe system called a loop field. Heat is thermally conducted to the water and is either drawn from or exhausted into the earth, depending on heating or cooling needs.

Closed-loop systems use an enclosed loop field (enclosed because the water in the pipes is not released to the environment) that relies on heat transfer across the pipes. A closed-loop system pumps water directly from an underground source such as a well or lake and either extracts heat from the water or adds heat to it before pumping it back to the source. A water-source heat pump is used to transfer heat to power an electricity generator or for direct heating and cooling. There are many commercially available water-source heat pumps for residential and commercial buildings. Water-source heat pumps transfer heat by either a water-to-air system, a water-to-water system, or a combination of the two. A water pump is also needed, to circulate water through or draw water from the loop field.

Geothermal energy production uses a minimal amount of electricity, primarily to run the water pump, and is a nonpolluting renewable energy source. The initial investment to design and install a geothermal system is currently higher than the cost of typical grid-supplied electricity systems. However, many commercially available systems report payback periods within three to five years, depending on specific project features, such as building size and location or depth of an available heat source. Geothermal energy systems are a promising option for many projects.

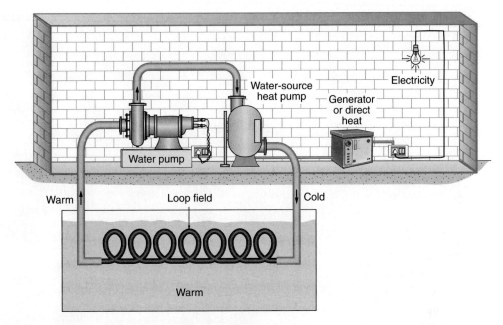

FIGURE 10.9 Shown here is a geothermal power system.

WIND TURBINE POWER SYSTEMS

Wind turbines convert wind forces into electricity. The phenomenon of wind is caused by uneven heating of the earth's atmosphere, surface irregularities, and the earth's rotation. Specific wind patterns are caused by particular types of terrain, water bodies, and vegetation. Wind energy is produced by converting the wind's kinetic energy into mechanical power that can be used for a specific function, such as pumping water (in which case the device is referred to as a "windmill"), or to run a generator that converts the mechanical energy into electricity. Wind turbines use the wind to turn blades that look like an airplane propeller; the blades turn a shaft connected to a generator to produce electricity. Small turbines (100 KW or less) can be used on-site to provide power for a building. Modern wind turbines are designed to maximize efficiency and have come a long way from the steel-truss–mounted windmills commonly seen on farms.

On-site turbines can be mounted on a traditional mast or attached to and integrated with the site features or the building. Mast- and building-mounted turbines are readily available from multiple commercial vendors. Many innovative designers are integrating wind turbines into the architecture of the building and suppliers are beginning to introduce architectural wind turbines that are mounted onto the building. One company, AeroVironment, Inc. (www.avinc.com), plans to make a compact wind turbine that sits on a building's parapet with an enhanced appearance that can be incorporated into the building's architecture. According to the company this design will also catch wind that travels up the side of the building, resulting in a potential 30% increase in energy production.

The amount of power generated by a wind turbine depends on climatic conditions such as wind speed. On-site wind-generated electricity is often combined in a "hybrid"

FIGURE 10.10 This windmill uses wind power to pump water from a well for farm irrigation.

system with an alternate electricity source. Wind speed and direction change with the time of day and the seasons and are difficult to predict. An alternate power source is often necessary to supplement wind-generated power during low wind periods. Wind power generation is most efficient for a site with relatively consistent wind patterns and speeds. To determine whether wind power generation is appropriate for a specific site, project engineers should first take wind speed and direction measurements over a sample period. Wind speed maps show the average wind information for specific areas and are available from turbine suppliers or any state's Energy Department, although they generally don't have information specific to a particular building site.

Turbine Configuration

Most modern turbines use one of two basic configurations:

- Turbines with a horizontal axis look similar to an airplane propeller and have either two or three blades.
- Turbines with a vertical axis look similar to an eggbeater and are commonly referred to as "Darrieus" turbines, after their inventor.

FIGURE 10.11 Modern wind turbines convert wind energy into electricity.

The amount of power produced by a wind turbine depends primarily on the wind speed, the turbine size, and the efficiency of the turbine's aerodynamics. Higher wind speeds will turn the turbines faster and produce more output. Many modern wind turbines can produce power with little wind (on the order of 6 mph), but are more efficient at higher wind speeds. Generally, larger turbines produce more power. Larger turbines are more efficient, providing more power with the same amount of wind, but they are more intrusive and may be difficult to integrate into the site or building design. Multiple smaller turbines can be mounted to roofs, walls, masts, or parapets and wired together to produce more power. Well-designed wind turbine systems will consider structural building factors such as weight and vibration along with noise isolation and location.

Wind turbines are commercially available in complete units that can be mounted to a free-standing mast or attached directly to a building and wired into the building's electrical system. The basic components of a wind turbine include the following:

- **The blades,** which, when the wind blows, turn a shaft that powers the generator.
- **The anemometer,** which measures and transmits wind speed data to the controller.
- **The brake,** which is used to stop the rotor in emergencies or for maintenance.
- **The controller,** which starts the system at low wind speeds (on the order of 6 mph) and shuts it off at high wind speeds (greater than about 55 mph) that can damage the wind turbine.
- **The gears,** which increase rotation speeds between rotors and generator.

- **The generator,** which converts mechanical energy from the shaft rotation into electricity.
- **The yaw vane and drive,** which measures the wind direction and orients the turbine to face into the wind as the wind direction changes.

Wind power is one of the largest renewable energy industries and grew by 28 percent worldwide in 2007, to reach about 95 GW. The annual capacity increased 40 percent in 2007 over its level in 2006.

HYBRID RENEWABLE ENERGY SYSTEMS

Hybrid systems use more than one type of renewable energy system to supply power and typically supply a greater proportion of a building's consumption needs than individual systems provide. Many hybrid systems take advantage of the strength and weaknesses of each energy system they use. For example, hybrid systems that use wind generators combined with PV panels take advantage of generally stronger wind currents during seasons with low light. Following are some examples of innovative hybrid systems:

- **The Augustenborg neighborhood,** Sweden, uses solar thermal and ground source heat to meet residents' needs with reduced emissions and socioeconomic sustainability.

FIGURE 10.12 Solar Fabrik photovoltaic panel plant is powered by PV panels and a vegetable oil cogeneration system.

- **The Vauban District in Freiburg,** Germany, features houses that are passively heated and cooled, resulting in a 90% reduction in energy use compared with conventional buildings.
- **Solar Fabrik in Freiburg,** Germany, is a zero-emission factory that makes photovoltaic panels and that is powered by PV and a vegetable oil cogenerator.

REFERENCES AND RESOURCES FOR FURTHER STUDY

1. AeroVironment, Inc. http://www.avinc.com.
2. Davidson, Sarah. National Survey Report of PV Power Applications in the United Kingdom., 2006.
3. Dunlop, Jim. *Photovoltaic Systems.* Homewood, IL: American Technical Publishers Inc., 2007.
4. GE Ecomagination: Wind Turbines. Video and facts about the first stage of the world's largest wind farm, which opened in Arklow, Ireland, in 2004. http://ge.ecomagination.com/site/showcase/arklow.html.
5. Geothermal Heat Pump Consortium. http://www.geoexchange.org/
6. National Association of State Energy Officials. http://www.NASEO.org.
7. Renewable Energy Policy Network for the 21st Century (REN21). *Renewables 2007 Global Status Report.* 2007. http://www.ren21.net/globalstatusreport/default.asp.
8. Southface Eco Office in Atlanta, Georgia. This is an example of a notable project using multiple innovative strategies to reduce the building's water use. http://www.southface.org.
9. U.S. Department of Energy. High Performance Buildings Database. www.eere.energy.gov/buildings/database/.
10. U.S. Environmental Protection Agency. Energy Star program. http://www.energystar.gov. The EPA certifies appliances that meet specific environmental criteria. Look for appliances that have an "Energy Star" rating.

Universities and institutes with a photovoltaic research department:

- The Center for Functional Nanomaterials at Brookhaven National Laboratory
- Solar Energy Laboratory at University of Southampton
- National Renewable Energy Laboratory (NREL)
- Energy & Environmental Technology Application Center at the College of Nanoscale Science and Engineering, SUNY, Albany
- Institut für Solare Energiesysteme ISE at the Fraunhofer Institute
- Energy Research Centre of the Netherlands (ECN)
- Imperial College London: Experimental Solid State Physics
- Instituto de Energía Solar at Universidad Politécnica de Madrid
- Centre for Renewable Energy Systems Technology at Loughborough University
- School of Photovoltaic and Renewable Energy Engineering at The University of New South Wales
- Centre for Sustainable Energy Systems at the Australian National University
- Ecole Polytechnique Fédérale de Lausanne (Dr. Graetzel invented dyesensitized cells here)
- Advanced Energy Systems at Helsinki University of Technology
- Institute of Materials Research at Salford University
- The Centre for Electronic Devices and Materials at Sheffield Hallam University
- The Solar Caliometry Lab at Queen's University
- Institute of Microtechnology at University of Neuchatel in Switzerland
- University of Konstanz
- Arizona State University Photovoltaic Testing Laboratory
- Institute of Energy Conversion at University of Delaware
- World Alliance for Decentralized Energy
- Florida Solar Energy Center at University of Central Florida
- Linz Institute for Organic Solar Cells (LIOS)

CHAPTER 11

Improving a Building's Material Use

OUTLINE

- Reusing Existing Building Stock
- Selecting Building Materials
- Building Materials That Have Recycled Content
- Material Transportation Impacts
- Building Materials That Are Made from Rapidly Renewable Resources
- Recycling Construction Waste
- References and Resources for Further Study

REUSING EXISTING BUILDING STOCK

Reusing existing structures, instead of building new ones, can conserve resources and reduce the waste and environmental impacts associated with construction, material manufacturing, and transportation. In many situations, an existing building can be renovated to provide a different use by utilizing design and construction methods that minimize the environmental impact of the project. This predesign phase activity commonly is completed by the owner with input from the architect and builder. The USGBC recommends the following strategies for reusing existing structures:

- Reuse as much of an existing structure as possible, including structural elements, the exterior envelope, and other building elements.
- Remove elements that pose contamination risks to building occupants like lead-based paint and asbestos.
- Upgrade building components, including windows, mechanical systems, and plumbing fixtures, to improve energy and water efficiency.

SELECTING BUILDING MATERIALS

Reusing building materials as much as possible will help reduce the demand on natural resources and decrease construction waste. Use salvaged and refurbished light fixtures, doors, cabinets, and flooring whenever appropriate and possible. Many retailers specialize in providing reused materials, and many architects and owners like the style of refurbished materials. Designers should involve the builder early in the material selection process to get input on the cost, quality, durability, and availability of recycled materials. The material selection process must take a holistic perspective, considering many factors:

- Harvesting of new materials.
- Social factors (such as stimulating local economies).
- Transportation of materials.
- Installation.
- Cost.
- Availability.

Sometimes, the consideration of these factors leads to unexpected conclusions. For example, even if a building material has been environmentally harvested and manufactured, its benefits may be offset by the pollution caused by transporting it from far away to the building site. It may be a stretch to consider a building material green if it has to be transported on a diesel ship halfway across the world. Many states have programs in place to help builders find recycled materials. One such program is the California Materials Exchange Program (CalMAX), which can be accessed at http://www.ciwmb.ca.gov/calmax/.

BUILDING MATERIALS THAT HAVE RECYCLED CONTENT

Many building products made from recycled materials are available. Using materials with recycled content helps the project avoid causing environmental impacts from the material extraction process for raw materials and can encourage more suppliers to use recycled materials in their manufacturing processes. Materials with post-consumer recycled content are made partially from waste produced by items people have already used, such as aluminum cans or newspaper. Materials with preconsumer recycled content are made partially from waste produced by the product-manufacturing process, like scraps from steel manufacturing. Many standard building products contain recycled content, including drywall, steel, and acoustical ceilings. The project team should consider factors such as durability, cost, and performance in choosing materials.

Many innovative strategies are being used to incorporate recycled materials such as the following into projects:

- Flooring made from recycled rubber.
- Cabinets made from recycled paper.
- Countertops made from recycled glass or porcelain from old toilets.
- Wall insulation made from old blue jeans that can provide thermal resistance properties that are similar to traditional insulation.
- Road base made from processed asphalt shingles.

MATERIAL TRANSPORTATION IMPACTS

Purchasing materials that are produced near the project site can reduce the environmental impact (and cost) of material transportation. The USGBC defines regional materials as those which are extracted, harvested, manufactured, or recovered within 500 miles of the project site.

BUILDING MATERIALS THAT ARE MADE FROM RAPIDLY RENEWABLE RESOURCES

Even if it is not possible or practical to use recycled materials, selecting materials that are made from a rapidly renewable source can reduce the depletion of nonrenewable and long-cycle–renewable materials. For example, some varieties of bamboo grow to full maturation in 3 to 5 years, while oak may take 20 or more years to mature enough to harvest and use. Bamboo can be used for floors, cabinets, veneers, plywood, structural members, scaffolding, and even concrete reinforcing. (Some bamboo has been shown to be as strong as, or stronger than, steel in tension.) Many traditional building materials can be replaced by rapidly renewable materials that provide a similar or superior performance:

- Wool for carpet pads and carpets.
- Cotton for insulation.
- Agrifibers like wheat stalks for wood building materials.
- Linoleum for counters and floors.
- Cork flooring.

The harvesting of wood for building materials like dimensional framing lumber, flooring, floor decks, and doors can significantly affect the environment if it is not completed in an environmentally sensitive manner. Many producers of wood products have adopted responsible forest management techniques to minimize the environmental impact of harvesting. Wood products purchased for a building project should be certified by an independent organization that uses measurable benchmarks including:

- governance;
- substantive technical standards;
- accreditation and auditing; and
- chain of custody and labeling.

The Forest Stewardship Council (FSC) is one such organization that has established criteria for environmentally responsible wood harvesting. The trademark of the FSC indicates that wood used to make the product comes from a forest which is well managed according to strict environmental, social and economic standards. The forest of origin has been independently inspected and evaluated according to the principles and criteria for

FIGURE 11.1 The FSC logo identifies products which contain wood from well managed forests certified by the Forest Stewardship Council.

Source: Courtesy Forest Stewardship Council © 1996 FSC A.C.

forest management agreed and approved by the Forest Stewardship Council. FSC is an international, non-profit association whose membership comprises environmental and social groups and progressive forest and wood retail companies working in partnership to improve forest management worldwide.

RECYCLING CONSTRUCTION WASTE

Studies indicate that demolition and construction waste makes up over 40% of the total waste stream that is introduced into landfills in many areas of the United States. Almost all construction waste is recyclable. Average new construction projects traditionally generate about 4 pounds of waste per square foot of building area. For example, constructing a 2,000-square foot house generates about 8,000 pounds of waste; the majority of which is typically disposed of in a landfill. Landfill operations have a negative impact on open space, generating water, air, and ground pollution. Much of the waste generated from a typical construction project can be recycled or significantly decreased with some proactive planning. Many contractors are finding that recycling construction waste can actually save money by reducing dumping fees and through the resale of recyclable waste.

Recyclable construction waste can be separated from nonrecyclable waste at the construction site or at the disposal area. In many communities, the disposal facility separates recyclable materials from comingled waste generated on-site. This allows the builder to dispose of all construction waste into one container. The container is hauled to the disposal facility where the recyclable materials are separated. This approach generally decreases the labor, space, and facilities necessary to separate construction waste on-site. However, most disposal facilities charge a fee for separating the recyclable materials from the comingled waste. Also, they generally guarantee only that half of the waste materials will be recycled.

Separating recyclable construction waste on-site can save the cost of this fee and guarantee that all of the appropriate waste is recycled. Once employees have become efficient at separating waste on-site, this approach can also reduce overall disposal costs. On-site separation requires the project manager to think ahead and formulate a specific plan to ensure that recycling facilities will be available and that all construction personnel are educated in implementing the plan. Workers need to be given specifically labeled storage bins for disposal of recyclable construction waste and must haul them to a recycling disposal area once the bins are full.

Common Construction Waste Items That Can Be Recycled:	
Asphalt	Plastic
Concrete	Ceiling Tiles
Cardboard	Carpet
Drywall	Asphalt Roofing
Metals	Wood
Glass	Aluminum

Successful implementation of this strategy is contingent on properly training all construction workers and subcontractors to recognize the types of materials that will be recycled and follow the appropriate procedures for separation. Initially, it may be challenging to change the habits of workers, but once workers become familiar with the separation procedures, they will become more efficient. The cost of on-site separation and disposal of recycled construction waste can be equal to or less than comingled waste separation strategies.

Reducing the amount of construction waste generated is a fundamental way to decrease pollution from construction activities. Consider the following recommendations to reduce the overall amount of waste materials generated:

- Optimize building layout to correspond with standard dimensions to reduce material waste from items such as wood framing and drywall.
- Develop detailed framing layouts to avoid waste when ordering lumber and steel, and make sure that installers use the appropriate materials.
- Store materials on a level surface under cover, in order to minimize waste-producing damage to the materials.
- Set aside lumber and plywood/OSB scraps that can be used later as fire blocking, spacers in header construction, and other purposes.
- Donate leftover building materials that are still usable to a charitable organization like Habitat for Humanity.
- Use clean wood scraps for blocking and/or chip them into landscaping mulch.
- Order materials in optimal dimensions and quantities to minimize waste.
- Set aside large drywall scraps during hanging for use as filler pieces in areas such as closets.
- Recycle drywall waste. Many facilities will recycle drywall waste into products such as textured wall sprays, acoustical coatings, gypsum stucco, fire barriers, or agricultural soil-conditioning products. (Gypsum, for example, can be applied to certain types of soil to make it more useful for farming.)
- Bury clean concrete chunks, old brick, broken blocks, and other masonry rubble on-site during foundation back-filling.
- Separate metals for recycling, such as copper piping, wire, aluminum siding, flashing and guttering, iron and steel banding from bundles, nails and fasteners, and rebar.
- Separate cardboard waste, bundle it, and store it in a dry place for recycling.
- Minimize the number of blueprints and reproductions generated during the design and construction process.
- Install leftover insulation in interior wall cavities or on top of installed attic insulation if it cannot be used on another job.

REFERENCES AND RESOURCES FOR FURTHER STUDY

1. California Materials Exchange Program (CalMAX), http://www.ciwmb.ca.gov/calmax/.
2. Forest Stewardship Council (FSC), http://www.fscus.org/.
3. Spiegel, Ross, and Dru Meadows. *Green Building Materials: A Guide to Project Selection and Specification*. New York: John Wiley and Sons, 1999.
4. United States Green Building Council (USGBC). *LEED for New Commercial Construction and Major Renovations, Version 2.2. Reference Guide*. Washington, DC, 2007.

CHAPTER 12

Improving a Building's Indoor Environment Quality

OUTLINE

- Ventilation Systems for Improved Indoor Air Quality
  - Measuring Indoor Air Quality
  - Mechanical Ventilation Systems
  - Passive Ventilation Systems
    - Cross-Ventilation
    - Stack Ventilation
  - Hybrid Integrated Ventilation Systems
- Methods for Improving Indoor Air Quality During Construction
  - Construction Air Quality Control Plan
  - Improving Air Quality During Construction
- References and Resources for Further Study

VENTILATION SYSTEMS FOR IMPROVED INDOOR AIR QUALITY

The quality of indoor air is directly related to the amount of ventilation in a building. A person's health, comfort, and well-being can be significantly affected by the quality of the indoor air. Buildings that do not have proper ventilation tend to be stuffy, smelly, and uncomfortable and may produce unhealthy environments. (Such buildings often are referred to as "sick" buildings.) Studies show that people who work in buildings with improved indoor air quality are more productive and have fewer sick days. Building ventilation systems are used to introduce fresh air from outdoors into the building while exhausting "old" air.

Measuring Indoor Air Quality

The amount of ventilation in a building affects the indoor air quality by introducing clean outside air into the building. Often the outside air is a different temperature from the temperature desired for air inside the building. For example, if the outside air is cold, then drawing it into the building for ventilation will introduce cold air into a heated space and cause discomfort for the occupants. The heating or cooling system will have to work harder to maintain a comfortable interior temperature. In order to mitigate the effects of ventilation on the heating and cooling system, most mechanical heating, ventilation, and cooling (HVAC) systems are designed to recirculate a significant portion of the indoor conditioned air. The recirculation of "old" air has a detrimental effect on the indoor air quality and should be monitored to verify that the air still meets minimum standards for building occupancy and use. It is virtually impossible to measure the actual amount of outside air included in the ventilation to an interior space. However, the amount of carbon dioxide (CO_2), a common byproduct in "old" air, can be measured to provide an indicator of air quality. CO_2 monitoring systems are commercially available and range from simple measuring devices to systems that can be integrated with other building systems to maximize the efficiency of the whole building.

Air quality is one of the factors in indoor environment quality. The overall quality of the indoor environment also takes into account factors such as temperature and humidity. Current technology using building management computer software and sophisticated measurement devices allows us to manage and control indoor environment quality. Some systems are simple devices that automatically open and close windows in accordance with measured level of indoor CO_2 and air temperature. Building management computer software allows input from measurement devices to guide the remote operation of building systems. Among the factors that can be manually or automatically controlled are heating, cooling, ventilation, the angle of exterior shade devices, and humidification. Measuring and controlling indoor environment quality is critical to maintaining a healthy and efficient building space.

Mechanical Ventilation Systems

Active (mechanical) ventilation systems use fans to draw fresh outside air into the building. Mechanical ventilation systems can provide consistent air flow to interior spaces and can be designed to maximize the indoor environment quality by controlling factors such as air speed, air quality, temperature, and humidity. High-efficiency filter systems can improve indoor air quality when they are integrated into the building's HVAC system. It is important for them to be highly efficient to avoid increasing the building's energy consumption. Well-designed mechanical ventilation systems are integrated with the heating and cooling systems. For example, outdoor air can be preheated before it is introduced into the building by integrating a heat exchanger into the HVAC system. The heat exchanger draws heat from the building's exhaust air. This heat is then used to preheat the outdoor air before it is introduced into the building's ventilation system.

Passive Ventilation Systems

Passive ventilation uses natural ventilation strategies to provide fresh outside air to the building without using mechanical systems. Well-designed passive ventilation systems are simple and require no maintenance or electrical power. Passive ventilation systems depend on outside air temperatures and weather patterns and may not be as efficient at different times of day or during different seasons. Humidity control of passive ventilation systems may be a challenge in areas that have extreme temperatures. The use of natural ventilation helps to decrease or eliminate the use of a building's mechanical ventilation systems, thereby saving energy. Passive ventilation system strategies are based primarily on the fundamental principles of cross- and stack ventilation.

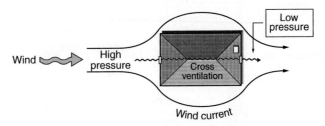

FIGURE 12.1 Cross-ventilation takes advantage of high- and low-pressure zones created by wind to draw fresh air into a building.

CROSS-VENTILATION. This system uses high- and low-pressure zones created by wind to draw fresh air into a building. Passive ventilation design strategically orients the building on the site in order to take advantage of prevailing wind patterns. The high- and low-pressure zones created as wind flows around a building can be used to provide natural airflow to the inside of the building. Ventilation devices are strategically located within, and integrated into, the building façade.

STACK VENTILATION. This system uses high- and low-pressure zones created by rising heat and the associated convection currents. As warm indoor air is vented from the building, it creates negative pressure in the building that can be used to draw outside air into the ventilation system. A passive ventilation system can move air by using heat generated by computers, lights, atria, and people. Outtake air vents are placed at the top of the building to exhaust warm air, while intake air vents in the lower levels of the building allow cooler air to enter.

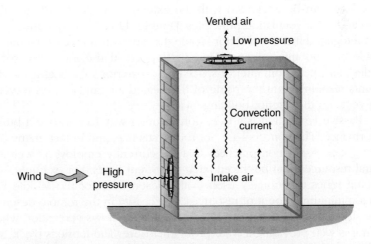

FIGURE 12.2 Stack ventilation takes advantage of high- and low-pressure zones created by rising heat and the associated convection currents.

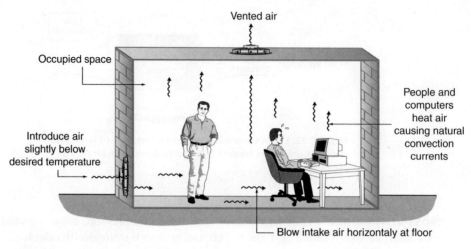

FIGURE 12.3 Displacement ventilation takes advantage of heat generated by people, computers, and other building uses to circulate air and provide ventilation.

Hybrid Integrated Ventilation Systems

Hybrid ventilation systems combine passive ventilation strategies with highly efficient mechanical ventilation and building systems. For example, displacement ventilation uses mechanically supplied ventilation that is introduced horizontally at the floor level. Supplied air is delivered at temperatures slightly lower than would be desired when the space is occupied. As the supply air moves horizontally across the floor, it is naturally heated by heat from interior building uses until it naturally rises. The resulting convective air current can move the air vertically to exhaust vents integrated with the ceiling system.

Raised floor air plenums (spaces that contain air) deliver air at the floor level and take advantage of the basic principle of displacement ventilation. A building's intake air ventilation system can be integrated with the exterior building façade to take advantage of both cross- and stack ventilation principles. Double skin façade systems have strategically located vents integrated into the exterior façade that take advantage of natural wind air flow to draw outside air into the building. This air trapped in the interstitial space between the two façades gains heat from interior space use or exterior solar heat gain. The natural convection currents resulting from the rising of the heated air can be used to ventilate interior spaces with vents on the interior building façade.

Passive ventilation is a free, nonpolluting way to improve a building's environmental performance. The concepts are not rocket science, and in fact, many of the passive heating, cooling, and ventilation practices that we currently employ, as well as the efficient use of natural resources, can be traced back to ancient times. For example, the tents that nomadic Bedouin tribes traditionally used in the desert areas of the Middle East were constructed with a minimum amount of resources, a necessity in the remote desert climate. The roofs of these structures were made of woven fabric that allows ventilation when it is dry but swells when it is wet, providing a waterproof enclosure. The fabric is black, which seems counterintuitive in a desert environment with temperatures reaching 120°F, as the color black absorbs heat. The "trick" is this: The black material's heat absorption results in a roof surface temperature that is greater than the ambient air temperature. This temperature differential causes a convection current at the roof surface. A clever roof shape takes advantage of the fact that heat rises, causing air currents to move along the roof surface that actually draw inside air through the roof fabric, providing natural ventilation and cooling. These types of basic concepts are being applied to modern buildings, making them more efficient and more comfortable and healthy to occupy.

METHODS FOR IMPROVING INDOOR AIR QUALITY DURING CONSTRUCTION

Construction activities can produce a significant amount of air pollution that can pose a health threat to construction workers and future building occupants. There are many strategies to avoid indoor air quality problems during construction and to mitigate long-term effects. Most of the methods for controlling indoor air quality during construction must be planned and implemented before construction begins.

Construction Air Quality Control Plan

An indoor air quality plan should be prepared and implemented before construction of the building. The plan should address the following factors:

- Methods for controlling the pollutant source.
- Mitigation of indoor air contaminant dispersion.
- Education for employees and subcontractors regarding the methods of pollution control.
- Ways to verify implementation of the plan.

During construction, it is necessary to provide proper ventilation to ensure that workers have a source of fresh outside air. Temporary fans can draw outside air into the building while filtering contaminants from the air before it is exhausted to the outside. If the building's permanent mechanical heating, venting, and cooling (HVAC) system will be used during construction, it must be protected from contaminants. Air intake vents should be covered with high-efficiency filters that meet the American Society of Heating, Refrigerating and Air-Conditioning Engineers (ASHRAE) MERV 8 rating minimum. Once construction is complete, the temporary filters can be removed and replaced with permanent ones. Specific building areas can also be isolated to reduce the area that is contaminated. Isolation is achieved by installing temporary barriers and depressurizing the region involved. (The air pressure difference between construction and clean areas helps to contain air contaminants.) It is generally not recommended that the building's HVAC system be used for demolition or other activities that generate significant indoor air pollutants.

Improving Air Quality During Construction

Proper cleanup and material storage during and after construction can significantly reduce the amount of air pollution generated by construction activities. Some good procedures to follow include

- Protecting porous materials from air contaminants before and after installation.
- Using a vacuum with a high-efficiency filter to clean up debris.
- Using wetting agents to control dust.

Many of the materials used during construction are odorous, irritating, and/or harmful to the health of the construction workers and building occupants. The storage and use of such materials should be specifically controlled to mitigate impacts on the indoor air quality. The amount of contaminants that a product emits is generally associated with the amount of volatile organic compounds (VOC) it contains. VOCs are carbon compounds that transform into a gaseous form when exposed to the environment. The gases produced by materials with high VOC levels can be harmful, and they interfere with the comfort of people in the building.

Many building material products are available with few or no VOCs, and they perform as well or better than traditional materials. Recommendations for specific VOC limits for different building materials can be found in the USGBC's LEED NC v2.2 Reference Guide available for purchase at http://www.usgbc.org/. The following materials generally contain VOCs and should be evaluated prior to use to determine whether reasonable substitutes exist:

- Adhesives and sealants.
- Paints and coatings.
- Carpeting and composition flooring.
- Composite wood products (ideally, these should be urea–formaldehyde free).

Once construction is complete, and before the building occupants move in, the building should be flushed out to remove any residual air pollution. This can be accomplished by blowing fresh outside air into the building while controlling temperature and humidity. The USGBC recommends supplying a total air volume of 14,000 cubic feet of outdoor air per square foot of building floor area while maintaining a temperature of at least 60 degrees Fahrenheit and a maximum relative humidity of 60%.

REFERENCES AND RESOURCES FOR FURTHER STUDY

1. Allen, Edward, and Joseph Iano. *Fundamentals of Building Construction* 4th ed. New York: John Wiley & Sons, 2004.
2. American Society of Heating, Refrigerating and Air-Conditioning Engineers (ASHRAE). http://www.ashrae.org.
3. Fisk, William J. *Health and Productivity Gains from Better Indoor Environments and their Relationship with Building Energy Efficiency.* U.S. Green Building Council, 2000.
4. Heschong, Lisa, et al. Daylighting in Schools: An Investigation, Into the Relationship Between Daylighting and Human Performance. Heschong Mahone Group, 1999.
5. Heschong, Lisa, et al. Skylighting and Retail Sales: An Investigation Into the Relationship Between Daylighting and Human Performance. Heschong Mahone Group, 1999.
6. James, J. P., and X. Yang. Emissions of Volatile Organic Compounds from Several Green and Non-Green Building Materials: A Comparison. *Indoor and Built Environment,* 2004
7. National Research Council. *Green Schools: Attributes for Health and Learning,* 2006. This report reviews and analyzes the results of studies on green schools and discusses the effects of green schools on student learning and teacher productivity.
8. Nicklas, Michael, and Gary Bailey. *Analysis of the Performance of Students in Daylit Schools,* Raleigh, NC: Innovative Design, 2006.
9. Seattle/King County Public Utilities Department. *Contractor's Guide,* 2002–2003.
10. U.S. Department of Energy. High Performance Buildings Database. http://www.eere.energy.gov/buildings/database/.
11. U.S. Green Building Council. *LEED-NC for New Construction Reference Guide Version 2.2* 1st ed. Washington, DC, 2007.

CHAPTER 13

Improving the Building Industry's Environmental Performance: The USGBC's LEED® Green Building Rating System™

OUTLINE

- Sustainability and the USGBC's LEED® Green Building Rating System
  - The U.S. Green Building Council
  - USGBC Guiding Principles
  - The LEED Green Building Rating System
  - The Future of LEED
- Becoming a LEED-Accredited Professional
- Building Certification
- References and Resources for Further Study

SUSTAINABILITY AND THE USGBC'S LEED® GREEN BUILDING RATING SYSTEM

Throughout history, the construction industry has attempted to improve the efficiency of buildings. In the 1970s, people began to realize that the building industry has a significant impact on the environment. Various groups and individuals attempted to influence the design and construction of buildings to make them more environmentally friendly. Unfortunately, this movement was generally associated with long hair and tie-dyed shirts and was not embraced by the building industry as a whole. During that time, significant worldwide decreases in energy costs further distracted the building industry from improving building efficiencies. Since then, the environmental movement that has most affected the building industry has been the U.S. Green Building Council.

The U.S. Green Building Council

The mission of the U.S. Green Building Council (USGBC), founded in 1998, is to transform the way buildings and communities are designed, built, and operated. This transformation is intended to foster an environmentally and socially responsible, healthy, and prosperous environment that improves the quality of life. USGBC is a nonprofit organization committed to expanding sustainable building practices. Its members include more than 15,500 organizations from across the building industry that are working to advance structures that are environmentally responsible, profitable, and healthy places to live and work. Member businesses include building owners and end-users, real estate developers, facility managers, architects, designers, engineers, general contractors, subcontractors, product and building system manufacturers, government agencies, and nonprofits.

USGBC Guiding Principles

- *Promote the Triple Bottom Line:* USGBC will pursue robust triple bottom line solutions that clarify and strengthen a healthy and dynamic balance between environmental, social, and economic prosperity.
- *Establish Leadership:* USGBC will take responsibility for both revolutionary and evolutionary leadership by championing societal models that achieve a more robust triple bottom line.
- *Reconcile Humanity with Nature:* USGBC will endeavor to create and restore harmony between human activities and natural systems.
- *Maintain Integrity:* USGBC will be guided by the precautionary principle in utilizing technical and scientific data to protect, preserve, and restore the health of the global environment, ecosystems, and species.
- *Ensure Inclusiveness:* USGBC will ensure inclusive, interdisciplinary, democratic decision-making with the objective of building understanding and shared commitments toward a greater common good.
- *Exhibit Transparency:* USGBC shall strive for honesty, openness, and transparency.

The LEED Green Building Rating System

The Leadership in Energy and Environmental Design (LEED) Green Building Rating System started by the USGBC is one of the first and most comprehensive tools for measuring and quantifying the environmental performance of buildings. The LEED Green Building Rating System is a voluntary, consensus-based national rating system for developing high-performance, sustainable buildings. LEED addresses all building types and emphasizes state-of-the-art strategies in five areas:

- Sustainable site development.
- Water savings.
- Energy efficiency.
- Materials and resources selection.
- Indoor environmental quality.

FIGURE 13.1 Use of the LEED
Certification Mark is authorized
by the USGBC LEED Department
once a project is LEED certified.

Source: Courtesy of the USGBC.

This model has also provided the building industry with a template and specific guidelines for designing and constructing greener buildings. The LEED rating system has now become the industry standard for designing and constructing highly efficient buildings, and its principles are being embraced across the building industry.

The USGBC has design and construction recommendations for many types of projects. Standard guidance documents include the following project categories:

- *New Commercial Construction and Major Renovations (LEED-NC)* is designed to guide and distinguish high-performance commercial and institutional projects.
- *Existing Buildings (LEED-EB)* focuses on operations and maintenance of existing buildings.
- *Commercial Interiors and Tenant Improvements (LEED-CI)* is a benchmark for the tenant improvement market that gives tenants and designers the power to make sustainable choices.
- *Core & Shell Development (LEED-CS)* aids designers, builders, developers, and new building owners in implementing sustainable designs for new core and shell construction.
- *Homes (LEED-H)* promotes the design and construction of high-performance green homes.
- *Neighborhood Development (LEED-ND)* integrates the principles of smart growth, urbanism, and green building into the first national program for neighborhood design.
- *Healthcare (LEED-HC)* promotes sustainable planning, design, and construction for high-performance healthcare facilities.
- *Retail (LEED-R)* recognizes the unique nature of retail design and construction projects and addresses their specific needs.
- *Schools (LEED-S)* addresses the specific and unique needs of K–12 schools.

FIGURE 13.2 USGBC allows builders of registered LEED for Homes projects that are actively seeking LEED certification to use this logo for on-site signage and for marketing materials specific to a project.

Source: Courtesy of the USGBC.

In this chapter, we focus on the overall concepts associated with sustainable design and building practices as outlined in the USGBC's *LEED for New Commercial Construction and Major Renovations* Version 2.2.

The Future of LEED

The USGBC is developing LEED Version 3.0, which will consolidate and align the different versions of LEED and incorporate recent advancements in science and technology into the guidelines. The new version is intended to implement a more flexible and adaptive program that will allow the USGBC to respond to the market's constantly evolving needs.

The new version of LEED will also consider more thoroughly the long-term implications of building projects through the end of a building's useful life and its deconstruction. A Life Cycle Analysis tool is being developed for inclusion in future LEED rating systems.

BECOMING A LEED-ACCREDITED PROFESSIONAL

The USGBC has a program for accrediting professionals who can demonstrate a familiarity with current sustainable building practices and technologies. Obtaining a LEED accreditation offers the following benefits to an individual or a project:

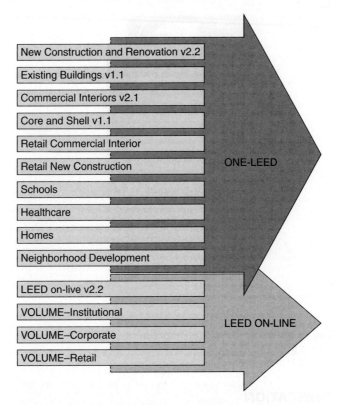

FIGURE 13.3 The USGBC is in the process of consolidating the current LEED Rating Systems into one program.

- LEED-accredited professionals (LEEDAP) are highly marketable to firms and clients.
- When a LEEDAP is involved in the design or construction of a building, the project earns a point toward LEED building certification.
- LEEDAPS are listed in the USGBC's Accredited Professionals Directory, which is available to all building professionals on the USGBC Web page.

In order to become a LEEDAP, an individual takes a comprehensive exam to demonstrate his or her knowledge of sustainable design, sustainable building practices, and the LEED rating system. While there are no prerequisites for taking the exam, the USGBC highly recommends the following qualifications:

- Experience in green building.
- Knowledge of the construction industry.
- Familiarity with the documentation process for LEED-certified projects.
- Knowledge of the intent, requirements, technologies, and strategies for each LEED Credit.
- Practical experience working with multiple design disciplines.

FIGURE 13.4   LEEDᴀᴘs strive to improve the environmental performance of building projects.

Source: Courtesy of the USGBC.

- Understanding of life-cycle costs and benefits of LEED.
- Familiarity with LEED resources and processes.

BUILDING CERTIFICATION

The heart of the USGBC's sustainable movement is its building certification program. Individual buildings that demonstrate high efficiency and little environmental impact can obtain a LEED certification. According to the USGBC,

> LEED Certification distinguishes building projects that have demonstrated a commitment to sustainability by meeting the highest performance standards. LEED was created to:
>
> - define "green building" by establishing a common standard of measurement;
> - promote integrated, whole-building design practices;
> - recognize environmental leadership in the building industry;
> - stimulate green competition;
> - raise consumer awareness of green building benefits;
> - transform the building market.
>
> LEED provides a complete framework for assessing building performance and meeting sustainability goals. Based on well-founded scientific standards, LEED emphasizes state of the art strategies for sustainable site development, water savings, energy efficiency, materials selection and indoor environmental quality. LEED recognizes achievements and promotes expertise in green building through a comprehensive system offering project certification, professional accreditation, training and practical resources.

There are many benefits to certifying a building project:

- Establish recognized leadership in the green building sector.
- Validate achievement through third-party review.
- Qualify for a growing array of state and local government incentives.
- Contribute to a growing green building knowledge base.
- Earn a LEED Certification plaque and certificate.
- Increase the value of the building and the potential for collecting higher rents from building occupants.

In order to qualify for a LEED rating, a building's sustainable design and construction practices must be documented using the LEED Letter Templates (available at the USGBC Web page at http//:www.usgbc.org/). Specific points are assigned to each LEED criterion that improves environmental performance. If the building achieves enough points, it wins a certification. Higher levels of environmental performance result in more points and a corresponding higher LEED rating. LEED ratings vary from the fundamental level of *Certified* to *Platinum*. The next step in improving environmental performance of buildings will be to create "living" buildings that offset their impacts on the environment. For now, LEED guidelines should be considered best practices and every project should obtain at least the LEED-certified level. The LEED Green Building Rating System is a benchmark for improving environmental performance.

The LEED rating system for New Commercial Construction and Major Renovations Version 2.2 (LEED-NC) addresses design and construction recommendations for incorporating sustainability into a project. The rating system is broken down into several general categories:

- Sustainable Sites.
- Water Efficiency.
- Energy and Atmosphere.
- Materials and Resources.
- Indoor Environmental Quality.
- Innovation in Design.

See Appendix A for detailed information about each category and the corresponding point qualifications and requirements for each.

The LEED-NC Green Building Rating System	
Certified	26–32 points
Silver	33–38 points
Gold	39–51 points
Platinum	52–69 points

REFERENCES AND RESOURCES FOR FURTHER STUDY

1. "LEED" and the related logo are trademarks owned by the U.S. Green Building Council and are used by permission.
2. "LEED Green Building Rating System" is a trademark owned by the U.S. Green Building Council and is used by permission.
3. All text, graphics, layout, and other elements of content referring to the U.S. Green Building Council's Leadership in Energy and Environmental Design (LEED®) Green Building Rating System™ are protected by copyright under both United States and foreign laws and are used by permission.
4. U.S. Green Building Council. *LEED for New Commercial Construction and Major Renovations* Version 2.2 Reference Guide. Washington, DC, 2007. http://www.usgbc.org.

CHAPTER 14

Improving Individual Environmental Performance: Becoming a LEED-Accredited Professional (LEEDAP)

OUTLINE

- LEED-Accredited Professional Programs and Qualifications
- Preparing for the LEED-NC Accreditation Exam
  - Focused Study Plan
  - USGBC Reference Guide
  - Time Line for LEED implementation
  - Additional Study Sources
- Exam Format
- Exam Specifications
  - Checklist: Knowledge of LEED Credit Intents and Requirements
  - Checklist: Coordinate Project and Team
  - Checklist: Implement LEED Process
  - Checklist: Verify, Participate In, and Perform Technical Analyses Required for LEED Credits
- Study Assessment Guide

LEED-ACCREDITED PROFESSIONAL PROGRAMS AND QUALIFICATIONS

The U.S. Green Building Council (USGBC) is a 501(c)(3) nonprofit whose members are leaders in the building industry, working to promote buildings and communities that are environmentally responsible, profitable, and healthy places to live and work. The Leadership in Energy and Environmental Design (LEED) Accredited Professional (AP) program sponsored by USGBC is for building professionals who can demonstrate a familiarity with current sustainable design and building practices and technologies.

A LEEDap candidate can select different areas of study for the exam, such as LEED for New Commercial Construction and Major Renovations (LEED-NC), LEED for Commercial Interiors (LEED-CI), and LEED for Existing Buildings (LEED-EB). Successful completion of any of the exam tracks leads to a LEEDap designation. This is a guide to the LEED-NC Version 2.2.

PREPARING FOR THE LEED-NC ACCREDITATION EXAM

Most candidates cannot complete this exam without specific preparation. Accreditation attests to the candidate's ability to facilitate the integrated design process, knowledge of the LEED Green Building Rating System, and understanding of the resources and processes involved with the project certification process. Thus, the best way for candidates to prepare is for them to understand the rating system requirements and processes and their practical application.

The main source of study material for the LEEDap exam is the LEED-NC v2.2 Reference Guide, supplemented by the references listed in the Study References section of this book. A list of resources, a description of the exam's content, and sample exam questions are available at http://www.gbci.org.

The LEEDap exam tests an applicant's knowledge of the LEED rating system and its implementation. The exam covers four general areas:

- Knowledge of LEED credit intents and requirements.
- Ability to coordinate the project and team.
- Ability to implement the LEED process.
- Capability of verifying, participating in, and performing technical analyses required for LEED credits.

A LEEDap candidate should have an extensive knowledge of each area.

Focused Study Plan

In order to understand the fundamental concepts of sustainable design and green building practices, LEEDap candidates must know how the overall green building process is implemented through the stages of design, construction, and operation. Candidates must (1) have specific knowledge of how strategies are implemented, (2) understand the strategies' intent, (3) know the requirements for measurement and documentation, and (4) be familiar with the potential technologies and strategies that are available. To truly embrace these concepts and prepare for this exam, candidates must study the whole process, not just any individual part of it, and must grasp the synergistic relationship among different strategies for improving the overall environmental performance of a building. Candidates must also be familiar with concepts from multiple sources across multiple disciplines.

To focus professional development goals towards improving environmental performance and prepare for the LEEDap exam, candidates should study all of the information in the Study References section of this book. Follow this step-by-step guide to help focus your study.

Step 1

- Download and study the Candidate Handbook from the Green Building Certification Institute Web page (http://www.gbci.org).

- Understand why this topic of study is important and how the process works in terms of the "big picture."

Step 2

- Read the LEED-NC v2.2 Reference Guide available for purchase from the USGBC (http://www.usgbc.org). As you read, highlight important areas for further study. Include areas that you don't fully understand or that are listed as important in the Exam Format section.
- Keep track of the important information from each section so that you can refer to it later. LEED-NC v2.2 Green Building Rating System Worksheet, included in Appendix B of this book, is a sample format for keeping track of important information.
- Get involved with some real-life projects that are using the LEED-NC v2.2 Green Building Rating System, and learn about the overall process and implementation of specific strategies.

Step 3

- Review the information listed in the Study References section.
- Study each reference and consider each of the items on the Checklist in the Exam Format Section. For example, know how to register a project for LEED certification online, manage the coordination of multiple job functions to achieve LEED certification, and complete the Credit Interpretations and Rulings (CIR) process.
- If you're confused or need help, ask your employer to provide training, and seek additional guidance from the USGBC.

STUDY HINT: Use the Exam Checklist in the Exam Format section of this book as your study guide. Once you understand each item, you're ready for the exam. Remember that the questions on this exam are randomly selected from a large pool of possible questions. This means that the questions on any exam may draw from any of the exam content areas. Candidates must understand every item on the Checklist.

Step 4

- Complete the Study Assessment Guide in Appendix C. Keep track of your time as you complete it. If you have a difficult time finishing the assessment in the amount of time allotted, or if you skipped straight to the assessment, go back to *Step 1.*
- Compare your answers with those given in the Answer Key in Appendix D. For each question, review the referenced material to make sure you understand the question and its answer.

STUDY HINT: Each of the answers in the Answer Key references the exam content area from the Checklist. Count the number of questions that you missed in each of the exam content areas. If you miss a significant number of questions in any of the exam content areas, go back and study material specific to that area again.

Step 5

- Read the LEED-NC v2.2 Reference Guide again, and concentrate specifically on the Credit Intent, Requirements, and Technologies and Strategies sections.

STUDY HINT: Recognize the difference between credits and points. For example, EA Credit 2 refers to On-Site Renewable Energy and is potentially worth 3 points.

- Review the outline you completed in *Step 2* and highlight areas that you don't remember or don't understand. If you didn't complete the outline, return to *Step 1.* Review the information in the Study Reference section in any area you're not sure about. If you don't understand the materials in an area, ask your employer to provide training, or contact the USGBC.

STUDY HINT: Candidates should be able to recall all the information included on the Green Building Rating System Worksheet in Appendix B for each credit.

- Continue to study the Reference Material until you have a fundamental understanding of each area on the Checklist. When you're ready to take the exam, check http://www.gbci.org to find out how to sign up for it.

Step 6
- Successfully complete the LEEDAP exam and add the designation to your business card.
- Implement what you have learned on all of your projects.

USGBC Reference Guide

The LEED-NC v2.2 Reference Guide is the primary study reference for this exam and includes information on the overall process and each of the LEED credit areas. The Reference Guide is organized into the following sections:

- Introduction.
- Sustainable Sites.
- Water Efficiency.
- Energy and Atmosphere.
- Materials and Resources.
- Indoor Environmental Quality.
- Innovation in Design.

The Reference Guide is further organized by specific credits. Candidates should be familiar with each credit and should specifically focus on the following aspects of each:

- Intent.
- Requirements.
- Process and timeline for implementation.
- Potential technologies and strategies.
- Definitions.
- Team profile and action items for decisions and implementation.

FOCUSED STUDY EXAMPLE

LEED-NC Reference Guide Reading Notes

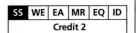

Sustainable Sites
Credit 2: Development Density and Community Connectivity

Intent:

Channel development to urban areas with existing infrastructure, protect greenfields, and preserve habitat and natural resources.

Requirements:

EB

OPTION 1—DEVELOPMENT DENSITY

Construct or renovate a building that is

• on a previously developed site, AND
• Is in a community with a minimum density of 60,000 net square feet per acre. Density calculation must include the project's area of disturbance and is based on a typical two-story downtown development.

OPTION 2—COMMUNITY CONNECTIVITY

Construct or renovate a building that
• Is on a previously developed site, AND
• Is within ½ mile of a residential zone or neighborhood with an average density of 10 units per net acre, AND
• Is within ½ mile of at least 10 basic services, such as food and clothing stores, AND
• Has pedestrian access between the building and the services.

Proximity is determined by drawing a ½-mile radius around the main building entrance on a site map and counting the services within that radius.

Potential Technologies & Strategies:

During the site selection process, give preference to urban sites that offer pedestrian access to a variety of services.

FIGURE 14.1   This figure explains how to interpret the LEED-NC Reference Guide.

Equations:

Sustainable Sites
Credit 2: Development Density and Community Connectivity

Equation 1:

$$\text{Development Density (SF/acre)} = \frac{\text{Gross Building Square Footage (SF)}}{\text{Project Site Area (acres)}}$$

Equation 2:

$$\text{Density Radius (LF)} = 3 \times \sqrt{\text{Property Area (acres)} \times 43{,}560 \text{ (SF/acre)]}}$$

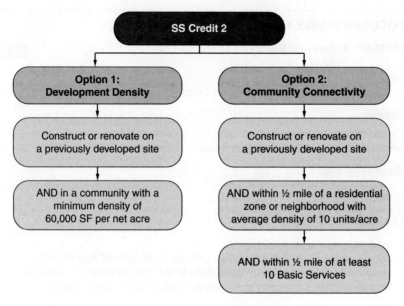

FIGURE 14.2 This flow diagram demonstrates the process for implementing SS Credit 2.

Time Line for LEED Implementation

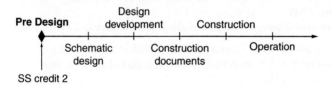

FIGURE 14.3 This figure is an example timeline for implementing SS Credit 2.

Action Items and Phase:

During the project's preconstruction phase, project managers should prepare to meet LEED objectives by

- Evaluating options for compliance.
- Compiling data and completing registration requirements at LEED-Online.

Additional Study Sources

The following sources provide more information about the LEED Rating System and its practical application (this list is also available from http://www.gbci.org):

- LEED Reference Guide: http://www.usgbc.org/store > Publications
- LEED certification process: http://www.usgbc.org/leed > Project Certification

- Project registration application: http://www.usgbc.org/leed > Project Certification > Registration
- LEED credit templates: http://www.usgbc.org/leed > LEED Resources > LEED Online Sample Credit Templates
- LEED-Online: http://www.usgbc.org/leed > Project Certification > LEED-Online
- Credit Interpretation Requests Process: http://www.usgbc.org/leed > Project Certification > Credit Interpretation Rulings > CIR and Ruling Process Guidelines

EXAM FORMAT

Each section of the exam is designed to test minimum competency in a specific area of the knowledge that a candidate must have in order to successfully facilitate the LEED-NC Version 2.2 certification process for a building.

Candidates for LEEDAP accreditation may not bring any references, notes, calculators, or other materials or tools into the exam. The exam consists of 80 questions administered in a 2-hour period, with up to 30 minutes to review the instructions for the exam. The questions are randomly selected from a large pool and are weighted differently.

EXAM SPECIFICATIONS

The specifications for each section of the LEED Professional Accreditation exam are organized to assess whether a candidate is capable of performing specific tasks and services. The following outline provides a general description of exam content areas.

Checklist: Knowledge of LEED Credit Intents and Requirements

- Understand and consistently apply LEED-related terms and definitions.
- Know about LEED credit intents requirements, submissions, technologies, and strategies for site, water, energy, materials, and Indoor Environmental Quality (IEQ) credit categories.
- Describe format and process for achieving innovation credits.

Checklist: Coordinate Project and Team

- Gather all project information and requirements to support the LEED process.
- Manage coordination of multiple job functions to achieve LEED certification.
- Identify standards that support LEED credits.
- Identify opportunities for integrated design and credit synergies to support LEED certification and explore systems integration opportunities.
- Identify critical path elements and schedule to implement the LEED process.
- Develop and implement a critical path for green building strategies.

Checklist: Implement LEED Process

- Select an appropriate LEED product for the project scope.
- Register the project for LEED certification online.
- Demonstrate knowledge of CIR process and resources.
- Manage the LEED documentation/certification process.
- Manage and complete letter templates.
- Draft and review innovation credits.

Checklist: Verify, Participate In, and Perform Technical Analyses Required for LEED Credits

- Verify that technical work products created by other team members comply with LEED standards.
- Participate in and guide the development of technical analyses with design professionals.
- Perform technical analyses to verify that building components comply with LEED requirements.

STUDY ASSESSMENT GUIDE

The Study Assessment Guide included in Appendix C is not a practice exam. It is an exercise that aims to help the student understand the fundamental concepts of sustainable design and green building and to apply them to every building project. Demonstrating LEED AP credentials is of secondary concern. Any individual who aspires to become a LEED AP must "practice what we preach" and demonstrate a commitment to improving the environmental performance of buildings. Passing the LEED AP exam is not enough to truly embrace the concepts of sustainability. A LEED professional also must embrace a personal commitment to improving environmental performance that is reflected in environmentally conscious personal decisions about consumption choices.

Example Study Assessment

1. EA Prerequisite 1: Fundamental Commissioning of the Building Energy Systems encourages owners to seek out qualified individuals to lead the commissioning process. Qualified individuals possess a high level of experience in which three of the following areas? (Choose 3)

    A. Energy systems design, installation and operation
    B. Long-term performance of the systems to be commissioned
    C. Hands-on field experience
    D. Energy systems automation control knowledge

    Answer: A, C, D

    LEED Manual Reference: EA Prerequisite 1: Fundamental Commissioning of the Building Energy Systems
    Study Area: Coordinate

Be familiar with the definition of a *qualified individual* who may document and implement each credit. EA Prerequisite 1: Fundamental Commissioning of the Building Energy Systems identifies qualified individuals as those who possess a high level of qualification in each of the following areas:

- Energy systems design, installation and operation
- Hands-on field experience
- Energy systems automation control knowledge

This Study Assessment Guide cannot determine whether a candidate is prepared to take the LEED-NC Accredited Professional Exam and should be used only to focus candidates' study. Since the exam questions are randomly selected for each candidate, it is impossible to generate a practice exam. The Study Assessment Guide is intended to help focus study and enhance understanding of fundamental sustainable design and green building practices. Each section of the Study Assessment Guide refers to a specific part of the LEED process. The Answer Key included in Appendix D includes answers to the Study Assessment Guide and References for further and more focused study.

LEED-NC Version 2.2 Registered Project Checklist

Project Name
City, State

Yes   ?   No

			Sustainable Sites	14 Points
Y			Prereq 1 **Construction Activity Pollution Prevention**	Required
			Credit 1 **Site Selection**	1
			Credit 2 **Development Density & Community Connectivity**	1
			Credit 3 **Brownfield Redevelopment**	1
			Credit 4.1 **Alternative Transportation**, Public Transportation Access	1
			Credit 4.2 **Alternative Transportation**, Bicycle Storage & Changing Rooms	1
			Credit 4.3 **Alternative Transportation**, Low-Emitting and Fuel-Efficient Vehicles	1
			Credit 4.4 **Alternative Transportation**, Parking Capacity	1
			Credit 5.1 **Site Development**, Protect or Restore Habitat	1
			Credit 5.2 **Site Development**, Maximize Open Space	1
			Credit 6.1 **Stormwater Design**, Quantity Control	1
			Credit 6.2 **Stormwater Design**, Quality Control	1
			Credit 7.1 **Heat Island Effect**, Non-Roof	1
			Credit 7.2 **Heat Island Effect**, Roof	1
			Credit 8 **Light Pollution Reduction**	1

Yes   ?   No

Water Efficiency   5 Points

Credit 1.1 **Water Efficient Landscaping**, Reduce by 50%   1
Credit 1.2 **Water Efficient Landscaping**, No Potable Use or No Irrigation 1
Credit 2   **Innovative Wastewater Technologies**   1
Credit 3.1 **Water Use Reduction**, 20% Reduction   1
Credit 3.2 **Water Use Reduction**, 30% Reduction   1

Yes   ?   No

Energy & Atmosphere   17 Points

Y   Prereq 1 **Fundamental Commissioning of the Building Energy Systems** Required
Y   Prereq 2 **Minimum Energy Performance**   Required
Y   Prereq 3 **Fundamental Refrigerant Management**   Required
         Credit 1 **Optimize Energy Performance**   1 to 10
         Credit 2 **On-Site Renewable Energy**   1 to 3
         Credit 3 **Enhanced Commissioning**   1
         Credit 4 **Enhanced Refrigerant Management**   1
         Credit 5 **Measurement & Verification**   1
         Credit 6 **Green Power**   1

Yes   ?   No

Materials & Resources   13 Points

Y   Prereq 1   **Storage & Collection of Recyclables**   Required
         Credit 1.1 **Building Reuse**, Maintain 75% of Existing Walls, Floors & Roof   1
         Credit 1.2 **Building Reuse**, Maintain 100% of Existing Walls, Floors & Roof 1
         Credit 1.3 **Building Reuse**, Maintain 50% of Interior Nonstructural Elements 1
         Credit 2.1 **Construction Waste Management**, Divert 50% from Disposal 1
         Credit 2.2 **Construction Waste Management**, Divert 75% from Disposal 1
         Credit 3.1 **Materials Reuse**, 5%   1
         Credit 3.2 **Materials Reuse**,10%   1
         Credit 4.1 **Recycled Content**, 10% (postconsumer + $\frac{1}{2}$ preconsumer)   1
         Credit 4.2 **Recycled Content**, 20% (postconsumer + $\frac{1}{2}$ preconsumer)   1
         Credit 5.1 **Regional Materials**, 10% Extracted, Processed & Manufactured Regionally 1
         Credit 5.2 **Regional Materials**, 20% Extracted, Processed & Manufactured Regionally 1

Yes   ?   No

			Credit 6 **Rapidly Renewable Materials**	1
			Credit 7 **Certified Wood**	1

Yes   ?   No

Indoor Environmental Quality   15 Points

Y			Prereq 1 **Minimum IAQ Performance**	Required
Y			Prereq 2 **Environmental Tobacco Smoke (ETS) Control**	Required
			Credit 1 **Outdoor Air Delivery Monitoring**	1
			Credit 2 **Increased Ventilation**	1
			Credit 3.1 **Construction IAQ Management Plan**, During Construction	1
			Credit 3.2 **Construction IAQ Management Plan**, Before Occupancy	1
			Credit 4.1 **Low-Emitting Materials**, Adhesives & Sealants	1
			Credit 4.2 **Low-Emitting Materials**, Paints & Coatings	1
			Credit 4.3 **Low-Emitting Materials**, Carpet Systems	1
			Credit 4.4 **Low-Emitting Materials**, Composite Wood & Agrifiber Products	1
			Credit 5 **Indoor Chemical & Pollutant Source Control**	1
			Credit 6.1 **Controllability of Systems**, Lighting	1
			Credit 6.2 **Controllability of Systems**, Thermal Comfort	1
			Credit 7.1 **Thermal Comfort**, Design	1
			Credit 7.2 **Thermal Comfort**, Verification	1
			Credit 8.1 **Daylight & Views**, Daylight 75% of Spaces	1
			Credit 8.2 **Daylight & Views**, Views for 90% of Spaces	1

Yes   ?   No

Innovation & Design Process   5 Points

			Credit 1.1 **Innovation in Design**: Provide Specific Title	1
			Credit 1.2 **Innovation in Design**: Provide Specific Title	1
			Credit 1.3 **Innovation in Design**: Provide Specific Title	1
			Credit 1.4 **Innovation in Design:** Provide Specific Title	1
			Credit 2 **LEED® Accredited Professional**	1

Yes   ?   No

Project Totals (pre-certification estimates)   69 Points

Certified 26–32 points   **Silver** 33–38 points   **Gold** 39–51 points   **Platinum** 52–69 points

Leadership in Energy and Environmental Design (LEED®) Green Building Rating System™ LEED Reference Guide Study Outline

Credit	Title	Points	Requirements	Intent	Calculation	Submittal	Timeline	Responsible Party	Exemplary Performance	Code
Sustainable Sites										
Credit 3 EXAMPLE	**Brownfield Redevelopment**	14 Points								
		1	Develop on a site documented as contaminated OR, on a site defined as a brownfield by a local, state or federal government agency.	Reduce pressure on undeveloped land by rehabilitating damaged sites.	None	Design	Pre-design	Civil Engineer	None	ASTM EPA
			Fill out this worksheet for each credit while you are studying the Reference Guide:							

LEED-NC Version 2.2 Professional Accreditation Study Assessment Guide

The LEED-NC exam covers four general areas:

- Knowledge of LEED credit intents and requirements.
- Ability to coordinate project and team.
- Ability to implement LEED process.
- Capacity to verify, participate in, and perform technical analyses required for LEED credits.

You may not bring any references, notes, calculators, or other tools or materials into the exam. The exam consists of 80 questions administered in a two-hour period with up to 30 minutes to review the instructions for the exam. The questions are randomly selected from a large pool of potential questions. Thus, a practice exam cannot provide an identical exam experience and should only be used to focus a candidate's further study.

This study assessment guide includes 80 multiple-choice questions and should be completed within two hours. Examinees may use one blank sheet of paper and a pencil for calculations.

Circle the most accurate answer(s) to each question.

1. The intent of SS Credit 4.4: Alternative Transportation: Parking Capacity is to reduce pollution and land development impacts from single-occupancy vehicle use. Which THREE of the following items may be used to verify compliance for a NONRESIDENTIAL project? (Choose 3)
   A. Size parking capacity to meet, but not exceed, minimum local zoning requirements
   B. Provide preferred parking for carpools or vanpools
   C. Provide shuttle services to mass transit
   D. Provide no new parking

2. WE Credit 1.1: Water-Efficient Landscaping: Reduce by 50% is intended to limit or eliminate the use for landscape irrigation of potable water or other natural surface or subsurface water resources available on or near the project site. Reductions shall be attributed to which of the following items?
   A. Plant species factor
   B. Irrigation efficiency
   C. Use of recycled wastewater
   D. Any combination of the above items
   E. None of the above

3. The intent of SS Credit 1: Site Selection is to avoid development of inappropriate sites and reduce the environmental impact from the location of a building on a site. Which of the following items may NOT be used as a criterion for compliance for developing buildings, hardscape, roads, or parking areas on any portion of a site?
   A. United States Department of Agriculture, United States Code of Federal Regulations, Title 7, Volume 6, Parts 400 to 699
   B. FEMA (Federal Emergency Management Agency) flood definition
   C. United States Code of Federal Regulations 40 CFR, Parts 230-233 and Part 22
   D. ASTM E1903-97 Phase II Environmental Site Assessment
   E. Federal or State threatened or endangered species lists

4. Long-term continuous measurement of boiler efficiency is required for which of the following credits?
   A. EA Prerequisite 1: Fundamental Commissioning of Building Energy Systems
   B. EA Credit 1: Optimize Energy Performance
   C. EA Credit 3: Enhanced Commissioning
   D. EA Credit 5: Measurement and Verification
   E. None of the above

5. SS Prerequisite 1: Construction Activity Pollution Prevention requires the creation and implementation of an Erosion and Sedimentation Control (ESC) Plan for all construction activities associated with the project. The ESC shall describe the measures implemented to accomplish all the following objectives EXCEPT:
   A. Prevent loss of soil during construction by stormwater runoff and/or wind erosion
   B. Protect topsoil by stockpiling for reuse
   C. Prevent sedimentation of storm sewer or receiving streams

  D. Evaluate the impacts of roof runoff
  E. Prevent polluting the air with dust and particulate matter

6. Which THREE of the following items may be used to verify site contamination in order to comply with SS Credit 3: Brownfield Redevelopment? (Choose 3)

   A. ASTM E1903-97 Phase II Environmental Site Assessment
   B. Local voluntary cleanup program
   C. Local, state, or federal government agency
   D. EPA Site Assessment Documentation Manual 2007

7. If your project reuses a portion of the building shell but not enough to qualify for MR Credit 1: Building Reuse, which of these holds true?

   A. You cannot use the credit for the reuse of the building shell under another credit elsewhere
   B. The building materials may be considered salvage materials and count towards that credit
   C. You may be able to claim the reuse activities under Construction Activity Pollution Prevention credit
   D. The materials may be considered regional materials

8. Which of the following is a nonprofit organization that promotes the manufacture and sale of environmentally responsible products used to verify compliance for EQ Credit 4.1, Low-Emitting Materials?

   A. Green Seal Organization
   B. Center for Resource Solutions
   C. EPA Energy Star Guidelines
   D. USGBC

9. You are constructing a project in an urban area, and you have access to a substantial amount of salvaged materials within a few hundred miles that you will incorporate into your project. Which of these will hold true?

   A. You may be able to claim 1 point for Materials Reuse
   B. You may be able to claim 1 point for Regional Materials (manufacturing point)
   C. You may be able to claim 1 point for Regional Materials (harvesting point)
   D. All of the above
   E. None of the above

10. EA Prerequisite 1: Fundamental Commissioning of the Building Energy Systems encourages owners to seek out qualified individuals to lead the commissioning process. Qualified individuals are identified as those who possess a high level of experience in which THREE of the following areas? (Choose 3)

    A. Energy systems design, installation, and operation
    B. Long-term performance of the systems to be commissioned
    C. Hands-on field experience
    D. Energy systems automation control knowledge

11. The Energy and Atmosphere category has three prerequisites and the Indoor Environmental Quality category has only two prerequisites. Identify THREE prerequisites for the EA category. (Choose 3)

    A. Minimum Energy Performance
    B. Environmental Tobacco Smoke (ETS) control
    C. Fundamental Refrigerant Management

    D. Fundamental Commissioning of the Building Energy Systems
    E. Ventilation Effectiveness

12. The required documentation for submission to the USGBC to achieve Water Efficient Landscaping Credits 1.1 and 1.2 includes which of the following items?

    A. The LEED letter template signed by the responsible party
    B. Calculations showing that the project uses 20% less water than the baseline fixture performance requirements of the Energy Policy Act of 1992
    C. Calculations showing that the site imperviousness is less than or equal to 50%
    D. A brief narrative describing the landscaping and irrigation design strategies

13. A commissioning agent used for Energy & Atmosphere Prerequisite 1, Fundamental Commissioning of the Building Energy Systems, would be responsible for which THREE of the following activities? (Choose 3)

    A. Creating a commissioning plan
    B. Designing the HVAC system with the mechanical subcontractor
    C. Integrating contractor-completed commissioning process into the construction documents
    D. Documenting the owner's project requirements

14. Which of the following strategies is NOT applicable to EA Credit 3, Enhanced Commissioning?

    A. Develop and utilize a commissioning plan
    B. Conduct a selective review of construction documents
    C. Conduct a selective review of contractor submissions
    D. Develop a recommissioning management manual

15. A renovation project is targeting "Silver" certification. The existing structure will leave the roof, exterior walls, and flooring in place and attempt to divert the rest of the waste from the landfill. This project should be registered under which LEED program?

    A. LEED-EB
    B. LEED-NC
    C. LEED-CI
    D. LEED-ND

16. Which of these Sections has the greatest number of possible points?

    A. Sustainable Sites
    B. Water Efficiency
    C. Energy and Atmosphere
    D. Materials and Resources
    E. Indoor Environmental Quality

17. Project teams interested in obtaining LEED Certification for their project must do which of the following?

    A. Upon completion of construction, evaluate whether project qualifies for LEED certification
    B. Check building codes to see if LEED compliance is required
    C. Register their project with the USGBC
    D. All of the above

18. Which of the following is true about having a LEED Accredited Professional on your project team?

    A. Is a prerequisite for Innovation and Design Process category
    B. Gets you 1 credit in Innovation and Design Process
    C. Is a requirement for the commissioning agent
    D. All of the above
    E. None of the above

19. Which of the following is NOT true regarding MR Credits 1.1, 1.2, and 1.3, Building Reuse?

    A. Significantly reduce construction waste
    B. Lower infrastructure costs
    C. Minimize habitat disturbance associated with greenfield site
    D. Often provide better-performing building envelopes

20. Which THREE of the following items are required to obtain a LEED certification for any building project? (Choose 3)

    A. Comply with the mandatory provisions of ASHRAE/IESNA Standard 90.1-2004
    B. Provide fundamental commissioning of the building energy system
    C. Comply with mandatory provisions of IESNA Energy Conservation LBNL-47493
    D. Provide Fundamental Refrigerant Management

21. An owner is designing a 250-unit condominium project and plans to register for LEED Certification. The design team intends to apply for Sustainable Sites Credit 4.2: Alternative Transportation, Bicycle Storage, and Changing Rooms. The owner expects to have 550 occupants and plans to provide secure bicycle spaces including bicycle racks, lockers, and storage rooms that are easily accessible by building occupants during all periods of the year, free of charge. How many secure bicycle spaces must the owner provide in order to obtain 1 point towards LEED Certification?

    A. 28
    B. 55
    C. 83
    D. None of the above

22. Which THREE of the following items are required to obtain a LEED certification for a new building? (Choose 3)

    A. Storage and Collection of Recyclables
    B. Controllability of Systems
    C. ETS Control
    D. Construction Activity Pollution Prevention

23. Your project intends to reuse a significant portion of the building shell, structure, and nonshell components. Which of the following statements is NOT true?

    A. You may be able to claim up to three credits for building reuse
    B. Building materials under this credit may be considered salvage materials
    C. Preserved materials also may be applied towards Construction Waste Management credit
    D. All of the above
    E. None of the above

24. An owner is designing a four-story office building that will be registered for a LEED Certification. The project team has estimated that the total cost of materials for the project (CSI Divisions 2-10) will be $13,587,000. The owner intends to install bamboo flooring that will cost $165,000, cotton batt insulation at $21,000, and wool carpeting at $95,000. Which of the following items will allow 1 point for MR Credit 6: Rapidly Renewable Materials?

A. Recycled brick veneer at $145,000

B. Linoleum flooring at $65,000

C. Wheatboard cabinetry at $55,000

D. Cork flooring at $53,000

25. Which TWO of the following will achieve points towards a project's LEED Certification? (Choose 2)

A. Exceeding the standards set in the LEED Credit requirements

B. Involving a LEED Accredited professional in guiding the project

C. Fundamental Refrigerant Management

D. Verifying installation, functional performance, training, and operation maintenance documentation

26. As a general rule of thumb, the Innovation in Design credit for exceptional performance is awarded for which TWO items? (Choose 2)

A. Doubling the credit requirements

B. Exceeding the credit requirement

C. Achieving the next incremental percentage threshold.

D. Complying with a higher standard of measure

27. The design team on a LEED Registered project is pursuing a unique strategy for improving ventilation effectiveness. It is unclear, on the basis of information provided in the Rating System and Reference Guide, whether or not the strategy satisfies the intent and requirements of a particular LEED credit and thus whether it might allow the project to earn much-needed points. After the team has read the Reference Guide with care, what is the next most important step the team should take toward determining the suitability of their design strategy for achieving the credit in question?

A. Search the list of LEED Certified projects for those having previously achieved the same credit

B. Send an inquiry to the TAG committee for the associated LEED credit category requesting clarification

C. Go to LEED-Online and complete a Credit Interpretation Request asking for clarification

D. Search the existing Credit Interpretation Rulings to see if there is one that addresses their question

28. The use of modulating photoelectric daylight sensors applies to which of the following areas?

A. Optimize Energy Performance

B. Daylight and Views

C. Light Pollution Reduction

D. Measurement and Verification

E. All of the above

F. None of the above

29. A design team is about to submit its project's LEED registration to the USGBC, but now some members of the team are questioning a proposed strategy for an ID Credit 1–1.4, Innovation in Design. Which of the following aspects should be considered as the team evaluates this strategy?

    A. Is there a measurable environmental benefit to this strategy?
    B. Is this an existing ID credit that has been previously accepted by the USGBC?
    C. Is this strategy already covered by an existing LEED Credit category?
    D. Is there a Credit Interpretation Request?
    E. All of the above
    F. None of the above

30. Your client's project will reuse an existing building shell, keeping the entire roof, floor, and all exterior walls; reduce the number of existing parking spaces to create an open space equal to 50% of the site area (excluding building footprint); and use native plants and adapted and low-water landscaping. This strategy will help achieve which THREE of the following credits? (Choose 3)

    A. SS Credit 5.1: Site Development
    B. MR Credit 1.1: Building Reuse
    C. WE Credit 1.1: Water Efficient Landscaping
    D. MR Credit 1.3: Building Reuse
    E. SS Credit 7.2: Heat Island Effect

31. Your new office building incorporates the use of modulating photoelectric daylight sensors and has installed PV cells that supply at least 2.5% of the building's total project annual energy cost. These strategies may contribute to a LEED Platinum rating by contributing to which TWO credits? (Choose 2)

    A. Optimize Energy Performance
    B. On-Site Renewable Energy
    C. Measurement and Verification
    D. Resource Reuse
    E. Green Power

32. The MEP Consultant has designed a system that will include a carbon dioxide monitoring system, mechanical ventilation, individual controls for temperature and humidity, and a permanent monitoring system. This strategy will help achieve at least three credits. Which TWO reference standards must be used to show compliance? (Choose 2)

    A. ASHRAE 62.1-2004, appendix C
    B. ASHRAE/IESNA 90.1-1999
    C. ASHRAE 55-2004
    E. The Energy Policy Act (EPAct) of 1992
    F. ASHRAE 129–97, Air-Change Effectiveness

33. You must perform a Soil/Climate analysis to determine appropriate landscape types for compliance with which of the following areas?

    A. Water Efficient Landscaping
    B. Water Use Reduction
    C. Landscape and Exterior Design to reduce heat island effect
    D. Reduced Site Disturbance

    E. All of the above

    F. None of the above

34. Strategies for reusing stormwater or greywater may be applied towards credit for which of the following areas?

    A. Water Use Reduction

    B. Water Efficient Landscaping

    C. Innovative Wastewater Technologies

    D. All of the above

    E. None of the above

35. The use of porous pavement systems with a high percentage of postindustrial waste applies to which of the following areas?

    A. Recycled Content

    B. Landscape & Exterior Design to reduce heat island effect

    C. Erosion and Sedimentation Control

    D. Resource Reuse

    E. Water Use Reduction

36. Credit Interpretation Request rulings provide which TWO of the following? (Choose 2)

    A. Responses to written requests for interpretation of credit requirements

    B. Determination of whether a particular strategy can be used to satisfy two different credits at once

    C. Precedents for interpretation of LEED credits regarding specific strategies and applications

    D. Definitive assurance that a particular method or strategy permitted on a previous project will be applicable to other projects in the future

37. Which of the following items is intended to reduce impacts from transportation and support the local economy?

    A. Site Selection

    B. Local/Regional Materials

    C. Alternative Transportation

    D. Renewable Energy

    E. All of the above

    F. None of the above

38. Building on sites already degraded, using native plants, designing a compact building, and preserving existing natural site amenities (strategies used together or separately) influence all of the following areas EXCEPT:

    A. Stormwater Management

    B. Reduced Site Disturbance

    C. Water Efficient Landscaping

    D. Site Selection

39. Maintenance of relative humidity levels between 30% and 60% will qualify for which of the following areas?

    A. Thermal Comfort

    B. Controllability of Systems

    C. Increase Ventilation Effectiveness

    D. Indoor Chemical and Pollutant Source Control

    E. All of the above

    F. None of the above

40. Who is included in the FTE?

    A. Full-time employees, visitors

    B. Full-time occupants, part-time occupants

    C. Full-time employees, part-time employees

    D. Full-time employees, part-time employees, customers

41. What would be useful in achieving WE Credit 1: Water Efficient Landscaping and SS Credit 5: Site Development?

    A. Vegetative swale

    B. Pervious parking

    C. Vegetative roof surface

    D. Rainwater reuse

42. What percent of potable water use must be reduced to achieve advanced WE Credit 3?

    A. 20%

    B. 30%

    C. 50%

    D. 75%

43. An owner wants to develop a previously developed site with imperviousness of 22% and a runoff of 200 fps (10,000 gallons per year). The postdevelopment imperviousness is 88%. What is the allowable amount of runoff rate (fps) and discharge (gallons) in order to achieve SS Credit 6?

    A. 200 fps and 10,000 gallons

    B. 500 fps and 20,000 gallons

    C. 1,000 fps and 50,000 gallons

    D. 1,200 fps and 100,000 gallons

44. A building has the following plumbing fixtures specified for bathrooms:

    5 low-flow toilets   1.1 gpf

    4 low-flow sinks   0.5 gpm

One of the building's occupants uses the toilet and sink three times a day each and washes her hands for 15 seconds each time. How many gallons of water does she use each day?

    A. 1.6 gal

    B. 3.7 gal

    C. 6.2 gal

    D. 8.4 gal

45. Which THREE of the following items must be included in the submission to register intent for a project's LEED-NC Certification? (Choose 3)

    A. Owner's name

    B. Advisor's name

    C. Name and location of job

    D. Name of subcontractors that are responsible for templates

    E. Certification goal of project

46. Which TWO of the following projects would most likely choose LEED-NC when registering for USGBC Certification? (Choose 2)

  A. New 40,000-sf building
  B. Remodel of a 40,000-sf building's HVAC system
  C. Remodel of a 40,000-sf building along with significant HVAC replacement
  D. Tenant Improvement of small building

47. Which THREE of the following activities are requirements of the commissioning authority to achieve EA Credit 3: Enhanced Commissioning? (Choose 3)

  A. Conduct one design review of the OPR and BOD
  B. Review contractor submissions
  C. Develop HVAC design calculations
  D. Review building operation within 10 months after substantial completion
  E. Install and test HVAC equipment

48. An owner is reusing existing walls, floors, and roofs, but not enough to achieve MR Credit 1: Maintain 75% of Existing Walls, Floors and Roofs. Which TWO of the following strategies may allow for potential points towards project certification? (Choose 2)

  A. MR Credit 1.1: Building Reuse
  B. MR Credit 5: Regional Materials
  C. MR Credit 6: Rapidly Renewable Materials
  D. MR Credit 2.1: Construction Waste Management
  E. MR Credit 7: Certified Wood

49. Which of the following items is NOT a strategy for achieving a point for EA Credit 2: On-Site Renewable Energy?

  A. Photovoltaic panel system
  B. Solar thermal water heat
  C. Geothermal water source heat pump and generator
  D. Low-impact hydro turbine
  E. Biomass generator

50. For a 1,000,000-sf office building (11.7 kWh/sf-yr, 58.5 kBtu/sf-yr) in New York ($0.1113/kWh, $0.00895/kBtu), determine how much renewable energy is required to achieve 1 point for EA Credit 2.

  A. $45,600
  B. $57,800
  C. $69,150
  D. $71,560
  E. $93,750

51. In order for a building to achieve points for EA Credit, its minimum energy performance must be established and compared against baseline calculations. Which of the following methods is used to calculate the unit cost value for baseline energy performance?

  A. Total average cost of energy divided by the total average annual expected energy consumption
  B. Annual expected energy consumption divided by the total average cost of energy
  C. Average monthly voltage drop from utility
  D. Energy Cost Budget Compliance Report

52. Which TWO of the following Credits require decision making by the contractor, design team, and owner? (Choose 2)

    A. EA Prerequisite 1: Fundamental Commissioning of the Building Energy System
    B. EA Prerequisite 3: Fundamental Refrigerant Management
    C. EA Credit 3: Enhanced Commissioning
    D. EA Credit 5: Measurement and Verification
    E. EA Credit 6: Green Power

53. Which of the following Credits is a construction phase submission?

    A. EA Credit 2: On-Site Renewable Energy
    B. EA Credit 4: Enhanced Refrigerant Management
    C. EA Credit 6: Green Power
    D. All of the above
    E. None of the above

54. ASHRAE Standard 55-2004 relates to which of the following areas?

    A. Thermal comfort conditions
    B. Building ventilation
    C. Individual thermal controls
    D. Air-cleaning devices

55. The Solar Reflectance Index (SRI) is a measurement of a material's ability to

    A. Reject solar heat
    B. Reflect solar light waves
    C. Solar surface radiation
    D. Absorb solar light waves

56. For a two-phase application, which THREE of the following items are included in the design phase submission? (Choose 3)

    A. EQ Prerequisite 1: Minimum IAQ Performance
    B. EQ Credit 2: Increased Ventilation
    C. EQ Credit 4: Low-Emitting Materials
    D. EQ Credit 6: Controllability of Systems

57. The Glazing Factor Calculation is an option to evaluate EQ Credit 8.1: Daylight and Views. Which THREE of the following are used to calculate the Glazing Factor? (Choose 3)

    A. Floor area
    B. T_{vis}
    C. Window height factor
    D. Glass SRI
    E. Gross building area

58. To satisfy EQ Credit 8: Daylight and Views, vision glazing must be placed between 2'-6" and 7'-6" above the finished floor for 90% of occupied areas. The square footage must meet which TWO of the following criteria? (Choose 2)

    A. In plan view, the area within sight lines from perimeter glazing
    B. In plan view, the area with direct sight lines to new or existing landscaping
    C. In section view, a direct sight line drawn from the area to perimeter vision glazing
    D. In section view, a direct sight line drawn from the area to a skylight (if interior wall obstruction only)

59. ASHRAE/IESNA 90.1-2004 is a standard used to evaluate which of the following items?

    A. Energy efficiency

    B. Ventilation effectiveness

    C. Thermal comfort

    D. Refrigerant use

60. A project needs to achieve more than 4 points for EA Credit 1: Optimize Energy Performance. Which of the following compliance path options would you recommend?

    A. Whole Building Energy Simulation

    B. Prescriptive Compliance Path–ASHRAE

    C. Prescriptive Compliance Path–Advanced Building Benchmark

    D. Energy Model—Test Case

61. Which THREE of the following activities must be completed by the project Commissioning Agent in order to achieve a point for EA Credit 3: Enhanced Commissioning? (Choose 3)

    A. Conduct at least one review of the OPR and BOD

    B. Review contractor submissions for systems being commissioned

    C. Develop systems manuals for future operating staff

    D. Verify that personnel training requirements are completed

    E. Review building operations within 10 months after substantial completion

62. Which THREE of the following options may be used to achieve EQ Credit 8: Daylight and Views? (Choose 3)

    A. Daylight Simulation Model

    B. Glazing Factor Calculation

    C. Building Model

    D. Daylight Measurement

63. Which of the following criteria relates to thermal comfort for building occupants?

    A. Air temperature

    B. Radiant temperature

    C. Relative humidity

    D. All of the above

    E. None of the above

64. Which THREE of the following items are required to comply with EQ Credit 5: Indoor Chemical and Pollutant Source Control for a mechanically ventilated building? (Choose 3)

    A. Provide 6-foot walk-off mat

    B. Create negative pressure in areas with chemicals present

    C. Test within 10 months after occupancy

    D. Install MERV-13 filters

65. Renewable energy sources used to achieve EA Credit 6: Green Power are defined by which of the following organizations?

    A. Center for Resource Solutions

    B. Department of Energy

    C. Renewable Energy Institute

    D. ASHRAE

66. The project manager has just been informed that the carpet that was supposed to be delivered to the project is no longer available. What should the advisor do in order to make sure the new carpet is compliant and no points are lost?

    A. Test carpet for VOC emission

    B. Verify nonfriable content

    C. Verify Green Label Plus Program certification

    D. Evaluate expected useful life

    E. Recalculate the CRI factor

67. A project design specifies doors made with a composite agrifiber core and wood veneer. Toward what Credit will this product most likely contribute?

    A. EQ Credit 4.4: Low-Emitting Materials: Composite Wood and Agrifiber Products

    B. ID Credit: Innovation in Design

    C. MR Credit 3: Materials Reuse

    D. MR Credit 6: Rapidly Renewable Materials

    E. MR Credit 7: Certified Wood

68. A project is trying to achieve the Rapidly Renewable Materials Credit with an overall cost of materials of $1,000,000. The project has already received shipments of $21,000 in renewable materials. Which item can be used in order to achieve one point?

    A. Cotton insulation for $4,500

    B. Agrifiber for $1,500

    C. Bamboo flooring for $2,000

    D. Linoleum flooring for $6,000

69. Which TWO of the following items should an administrator consider when he or she evaluates a previously developed site for redevelopment? (Choose 2)

    A. Located within ¹/₂-mile radius from a residential area

    B. Community density

    C. Elevation relative to 100-year flood

    D. Distance to closest body of water

70. In order for a building to achieve SS Credit 4.2: Alternate Transportation, which of the following items must a commercial or institutional building provide?

    A. Bicycle storage for at least 5% of building occupants

    B. Covered bicycle storage for 15% of building occupants

    C. Shower facilities for at least 5% of FTE occupants

    D. All of the above

    E. None of the above

71. A project requires 2 tons of concrete which is made up of 100 lbs of fly ash. What percent of the fly ash can apply to MR Credit 4: Recycled Content?

    A. 5%

    B. 10%

    C. 15% preconsumer/25% postconsumer

    D. 25% preconsumer/15% postconsumer

72. EQ Credit 4.4: Low-Emitting Materials does not allow the use of urea-formaldehyde resins and applies to which TWO of the following items? (Choose 2)

A. Interior doors
B. Furniture
C. Concrete slab formwork
D. Composite wood floor

73. Which of the following is NOT a likely strategy for achieving SS Prerequisite 1: Construction Activity Pollution Prevention?

A. Temporary seeding
B. Mulching
C. Earth dike installation
D. Sediment basin installation
E. All of the above
F. None of the above

74. Which TWO items will most likely contribute towards credit for MR Credit 2: Construction Waste Management? (Choose 2)

A. Foundation excavation soil that can't be reused on-site
B. Excess paint
C. Demolished copper water pipe
D. Land-clearing debris
E. Lumber scraps

75. Before submitting a Credit Interpretation Request (CIR), the applicant must

A. Review the intent of the credit or prerequisite in question and self-evaluate whether the project meets this intent
B. Consult the LEED Reference Guide for more detailed explanation, instructions, calculations, and guidance
C. Review the CIR pages to see whether the same inquiry has been answered previously or whether there are relevant CIRs that can help him or her deduce the answer
D. Contact LEED customer service to look into the request and confirm that it warrants a new CIR
E. All of the above

76. A project intends to use the flush-out compliance path to achieve EQ Credit 3.2: Construction IAQ Management Plan. Which of the following activities can be done concurrently with the flush-out?

A. Countertop installation using low-VOC glue
B. Final cleaning
C. Final HVAC testing and balancing
D. All of the above
E. None of the above

77. Which of the following project team members is most likely to carry decision-making responsibility for SS Credit 5.1: Site Development, Protect or Restore Habitat?

A. Contractor
B. Owner
C. Design team
D. All of the above
E. None of the above

78. What is the general rule of thumb for achieving ID Credits for exemplary performance? (Choose 2)

  A. Doubling the credit requirements
  B. Achieving the next incremental percentage threshold
  C. Increasing performance by a minimum of 50%
  D. Exceeding baseline calculations by 50%

79. The contractor is responsible for documenting which TWO of the following items during the project's construction phase? (Choose 2)

  A. Provide MSDSs with paint submissions
  B. Provide energy modeling calculations
  C. Provide site photometric plans
  D. Provide calculations for waste diverted from landfill

80. The owner of a 50-year-old office building has decided to replace the mechanical system to improve energy efficiency and is implementing green procurement and operations policies. Which LEED rating system is most applicable to this project?

  A. LEED-CI
  B. LEED-EB
  C. LEED-CS
  D. LEED-NC
  E. LEED-ND

Study Assessment Guide Answer Key and Focused Study Resources

LEED-NC VERSION 2.2 PROFESSIONAL ACCREDITATION EXAM

This Study Assessment Guide is intended to help focus readers' study and understanding of fundamental sustainable design and green building practices. Each section of the Study Assessment Guide refers to specific areas of the LEED process. To truly embrace the concepts of sustainability, readers not only must understand the fundamental ideas of sustainability, but also must implement them in their personal and professional activities. The fundamental body of knowledge possessed by LEED-accredited professionals is summarized into four fundamental areas:

1. *Credit knowledge* of LEED-NC credit intents and requirements
2. *Coordinate* project and team
3. *Implement* LEED-NC process
4. *Verify,* participate in, and perform technical analysis required for LEED-NC credits

This Answer Key includes answers to the Study Assessment Guide and references for further study. Each answer includes the general fundamental area for each area listed above. Count the number of questions that you miss in each area to see whether you need more focus in any one area. This Study Assessment Guide cannot determine whether a candidate is prepared to take the LEED-NC Accredited Professional Exam and should be used only to focus study. Since the exam questions are randomly selected for each candidate, it is impossible to create a representative practice exam.

STUDY ASSESSMENT GUIDE ANSWER KEY

1. The intent of SS Credit 4.4: Alternative Transportation: Parking Capacity is to reduce pollution and land development impacts from single-occupancy vehicle use. Which THREE

of the following items may be used to verify compliance for a NONRESIDENTIAL project? (Choose 3)

A. Size parking capacity to meet, but not exceed, minimum local zoning requirements
B. Provide preferred parking for carpools or vanpools
C. Provide shuttle services to mass transit
D. Provide no new parking

Answer: A, B, D

LEED Guide Reference: SS Credit 4.4: Alternative Transportation: Parking Capacity
Study Area: Credit Knowledge

Review the specific requirements and options to document and verify compliance for each credit. SS Credit 4.4: Alternative Transportation: Parking Capacity allows several *options* for compliance depending on the type of project:

OPTION 1—NONRESIDENTIAL

Size parking capacity to meet, but not exceed, minimum local zoning requirements AND provide preferred parking for carpools or vanpools for 5% of the total provided parking spaces.

OPTION 2—NONRESIDENTIAL

For projects that provide parking for less than 5% of FTE building occupants: provide marked preferred parking for carpools or vanpools for 5% of total provided parking spaces.

OPTION 3—RESIDENTIAL

Size parking capacity to remain within minimum local zoning requirements AND provide infrastructure and support programs to facilitate shared vehicle usage such as carpool drop-off areas, designated parking for vanpools, or car-share services, ride boards, and shuttle services to mass transit.

OPTION 4—ALL

Provide no new parking.

2. WE Credit 1.1: Water-Efficient Landscaping: Reduce by 50% is intended to limit or eliminate the use for landscape irrigation of potable water or other natural surface or subsurface water resources available on or near the project site. Reductions shall be attributed to which of the following items?

A. Plant species factor
B. Irrigation efficiency
C. Use of recycled wastewater

D. Any combination of the above items

E. None of the above

Answer: D

LEED Guide Reference: WE Credit 1.1: Water Efficient Landscaping

Study Area: Credit Knowledge

Refer to WE section in the Guide. WE Credit 1.1: Water Efficient Landscaping allows reductions attributed to any combination of the following items:

- Plant species factor.
- Irrigation efficiency.
- Use of captured rainwater.
- Use of recycled wastewater.
- Use of water treated and conveyed by a public agency specifically for nonpotable uses.

Be prepared to do calculations for each section where required and know what goes into the calculations. WE Credit 1.1 is calculated using:

- Landscape Coefficient (KL)–volume of water lost via evapotranspiration based on species climate and density.
- Species Factor (ks)–water needs of plant species.
- Density Factor (kd)–number of plants and total leaf area.
- Microclimate Factor (kmc)–environmental conditions: temp, wind, and humidity.

3. The intent of SS Credit 1: Site Selection is to avoid development of inappropriate sites and reduce the environmental impact from the location of a building on a site. Which of the following items may NOT be used as a criterion for compliance for developing buildings, hardscape, roads, or parking areas on any portion of a site?

   A. United States Department of Agriculture, United States Code of Federal Regulations, Title 7, Volume 6, Parts 400 to 699

   B. FEMA (Federal Emergency Management Agency) flood definition

   C. United States Code of Federal Regulations 40 CFR, Parts 230–233 and Part 22

   D. ASTM E1903-97 Phase II Environmental Site Assessment

   E. Federal or State threatened or endangered species lists

Answer: D

LEED Guide Reference: SS Credit 1: Site Selection

Study Area: Coordinate

Review the required standards for each credit. The reference standards in the Guide are for your information. However, you should be familiar with every required specific standard that is used to evaluate compliance for every credit. In order to earn SS Credit 1: Site Selection, you cannot develop buildings, hardscape, roads, or parking areas on portions of sites that meet any one of the following criteria:

- Prime farmland - United States Department of Agriculture in the United States Code of Federal Regulations, Title 7, Volume 6.

- Previously undeveloped land whose elevation is lower than 5 feet above the elevation of the 100-year flood–FEMA (Federal Emergency Management Agency).
- Land that is specifically identified as habitat for any species on Federal or State threatened or endangered lists.
- Land that is within 100 feet of any wetlands as defined by United States Code of Federal Regulations 40 CFR.
- Previously undeveloped land that is within 50 feet of a water body, defined as seas, lakes, rivers, streams, and tributaries that support or could support fish, recreation, or industrial use–Clean Water Act.
- Public parkland, unless land of equal or greater value as parkland is accepted in trade by the public landowner (Park Authority projects are exempt).

4. Long-term continuous measurement of boiler efficiency is required for which of the following credits?

A. EA Prerequisite 1: Fundamental Commissioning of Building Energy Systems
B. EA Credit 1: Optimize Energy Performance
C. EA Credit 3: Enhanced Commissioning
D. EA Credit 5: Measurement and Verification
E. None of the above

Answer: E

LEED Guide Reference: Credits

Study Area: Verify

Refer to the Guide and review the specific compliance process and requirements for each credit. For example,

- EA Prerequisite 1: Fundamental Commissioning of Building Energy Systems and EA Credit 1: Optimize Energy Performance is completed during the design and construction phase.
- EA Credit 3: Enhanced Commissioning requires staff training and involvement in reviewing building operation within 10 months after the building's completion.
- EA Credit 5: Measurement and Verification requires a Measurement and Verification Plan to evaluate energy system performance but only has to cover a 1-year period.

5. SS Prerequisite 1: Construction Activity Pollution Prevention requires the creation and implementation of an Erosion and Sedimentation Control (ESC) Plan for all construction activities associated with the project. The ESC shall describe the measures implemented to accomplish all the following objectives EXCEPT:

A. Prevent loss of soil during construction by stormwater runoff and/or wind erosion
B. Protect topsoil by stockpiling for reuse
C. Prevent sedimentation of storm sewer or receiving streams
D. Evaluate the impacts of roof runoff
E. Prevent polluting the air with dust and particulate matter

Answer: E

LEED Guide Reference: SS Prerequisite 1: Construction Activity Pollution Prevention

Study Area: Credit Knowledge, Coordinate

Refer to the Guide, review the objectives for each credit, and understand when they are completed in the overall process. The submission process is divided between Construction Activities and Design Activities. For example, SS Prerequisite 1: Construction Activity Pollution Prevention has the following objectives that are accomplished during the *construction phase*:

- Prevent loss of soil during construction by stormwater runoff and/or wind erosion.
- Protect topsoil by stockpiling for reuse.
- Prevent sedimentation of storm sewer or receiving streams.
- Prevent polluting the air with dust and particulate matter.

The impacts of roof runoff are evaluated during the *design phase*.

6. Which THREE of the following items may be used to verify site contamination in order to comply with SS Credit 3: Brownfield Redevelopment? (Choose 3)

   A. ASTM E1903-97 Phase II Environmental Site Assessment
   B. Local voluntary cleanup program
   C. Local, state, or federal government agency
   D. EPA Site Assessment Documentation Manual 2007

Answer: A, B, C

LEED Guide Reference: SS Credit 3: Brownfield Redevelopment

Study Area: Coordinate

Refer to the Guide and review the required standards for each credit. See Question 3 for a more detailed description. The required standards often are not intuitive because they don't sound very official. For example, SS Credit 3: Brownfield Redevelopment allows use of a *Voluntary Cleanup Program* to verify compliance with *Site Contamination* requirements.

7. If your project reuses a portion of the building shell but not enough to qualify for MR Credit 1: Building Reuse, which of these holds true?

   A. You cannot use the credit for the reuse of the building shell under another credit elsewhere
   B. The building materials may be considered salvage materials and count towards that credit
   C. You may be able to claim the reuse activities under Construction Activity Pollution Prevention credit
   D. The materials may be considered regional materials

Answer: B

LEED Guide Reference: Credits

Study Area: Coordinate

Review each credit and understand how they affect one another. Often a strategy can potentially relate to multiple credits depending on the specific requirements for each.

For example it's common to assume that reusing a portion of the building shell will apply to MR Credit 1: Building Reuse, but it may also be applied to MR Credit 3.1, which considers material reuse, or MR Credit 2.1, which attempts to divert construction waste from landfills.

8. Which of the following is a nonprofit organization that promotes the manufacture and sale of environmentally responsible products used to verify compliance for EQ Credit 4.1, Low-Emitting Materials?

   A. Green Seal Organization
   B. Center for Resource Solutions
   C. EPA Energy Star Guidelines
   D. USGBC

   Answer: A

   LEED Guide Reference: EQ Credit 4.1, Low-Emitting Materials

   Study Area: Coordinate

   Refer to the Guide and review the organizations that are used to verify compliance for each credit. The *Green Seal Organization* is used to verify the *Aerosol Adhesives* requirement associated with EQ Credit 4.1, Low-Emitting Materials.

9. You are constructing a project in an urban area, and you have access to a substantial amount of salvaged materials within a few hundred miles that you will incorporate into your project. Which of these will hold true?

   A. You may be able to claim 1 point for Materials Reuse
   B. You may be able to claim 1 point for Regional Materials (manufacturing point)
   C. You may be able to claim 1 point for Regional Materials (harvesting point)
   D. All of the above
   E. None of the above

   Answer: D

   LEED Guide Reference: Credits

   Study Area: Coordinate

   Refer to the Guide and review the requirements for achieving points for each credit. Be prepared to evaluate strategies that incorporate multiple points and how each is evaluated. For example, MR Credit 5: Regional Materials has multiple related points that also may apply to MR Credit 3: Material Reuse. Design should evaluate the extraction and manufacturing process for each material. Using reclaimed and locally harvested and manufactured materials reduces environmental impacts.

10. EA Prerequisite 1: Fundamental Commissioning of the Building Energy Systems encourages owners to seek out qualified individuals to lead the commissioning process. Qualified individuals are identified as those who possess a high level of experience in which THREE of the following areas? (Choose 3)

    A. Energy systems design, installation, and operation
    B. Long-term performance of the systems to be commissioned

C. Hands-on field experience

D. Energy systems automation control knowledge

Answer: A, C, D

LEED Guide Reference: EA Prerequisite 1: Fundamental Commissioning of the Building Energy Systems

Study Area: Coordinate

Be familiar with the requirements for *qualified individuals* required for documentation and implementation of each credit. For example, EA Prerequisite 1: Fundamental Commissioning of the Building Energy Systems identifies *Qualified Individuals* as those who possess a high level of qualifications in each of the following areas:

• Energy systems design, installation, and operation.

• Hands-on field experience.

• Energy systems automation control knowledge.

11. The Energy and Atmosphere category has three prerequisites and the Indoor Environmental Quality category has only two prerequisites. Identify THREE prerequisites for the EA category. (Choose 3)

   A. Minimum Energy Performance

   B. Environmental Tobacco Smoke (ETS) control

   C. Fundamental Refrigerant Management

   D. Fundamental Commissioning of the Building Energy Systems

   E. Ventilation Effectiveness

   Answer: A, C, D

   LEED Guide Reference: Credits

   Study Area: Coordinate

   Refer to Guide and review prerequisite *requirements* for each category. The Energy and Atmosphere category includes three required prerequisites:

   • Minimum Energy Performance.

   • Fundamental Refrigerant Management.

   • Fundamental Commissioning of the Building Energy Systems.

   Indoor Environmental Quality category includes two required prerequisites:

   • Environmental Tobacco Smoke (ETS) control.

   • Ventilation Effectiveness.

12. The required documentation for submission to the USGBC to achieve Water Efficient Landscaping Credits 1.1 and 1.2 includes which of the following items?

   A. The LEED letter template signed by the responsible party

   B. Calculations showing that the project uses 20% less water than the baseline fixture performance requirements of the Energy Policy Act of 1992

C. Calculations showing that the site imperviousness is less than or equal to 50%

D. A brief narrative describing the landscaping and irrigation design strategies

Answer: D

LEED Guide Reference: WE Credit 1: Water Efficient Landscaping

Study Area: Implement

Refer to the Guide and review the submission documentation requirements for each credit. For example the *required documentation* for submission to the USGBC to achieve Water Efficient Landscaping Credits 1.1 and 1.2 includes the following:

- Narrative describing the landscaping and irrigation design strategies.
- The project's calculated baseline Total Water Applied (TWA).
- The project's calculated design case TWA.
- Total nonpotable water supply available for irrigation.

13. A commissioning agent used for Energy & Atmosphere Prerequisite 1, Fundamental Commissioning of the Building Energy Systems, would be responsible for which THREE of the following activities? (Choose 3)

   A. Creating a commissioning plan

   B. Designing the HVAC system with the mechanical subcontractor

   C. Integrating contractor-completed commissioning process into the construction documents

   D. Documenting the owner's project requirements

   Answer: A, C, D

   LEED Guide Reference: Prerequisite 1, Fundamental Commissioning of the Building Energy Systems

   Study Area: Verify

   Refer to the Guide and review the roles and responsibility of the commissioning agent associated with the Energy and Atmosphere section. The Energy & Atmosphere Prerequisite 1, Fundamental Commissioning of the Building Energy Systems commissioning agent does not participate in the design of the system, but does review designs.

14. Which of the following strategies is NOT applicable to EA Credit 3, Enhanced Commissioning?

   A. Develop and utilize a commissioning plan

   B. Conduct a selective review of construction documents

   C. Conduct a selective review of contractor submissions

   D. Develop a recommissioning management manual

   Answer: A

   LEED Guide Reference: EA Credit 3, Enhanced Commissioning

   Study Area: Verify

The commissioning plan is a prerequisite in Fundamental commissioning.

15. A renovation project is targeting "Silver" certification. The existing structure will leave the roof, exterior walls, and flooring in place and attempt to divert the rest of the waste from the landfill. This project should be registered under which LEED program?

   A. LEED-EB
   B. LEED-NC
   C. LEED-CI
   D. LEED-ND

   Answer: B

   LEED Guide Reference: Introduction, http://www.usgbc.org

   Study Area: Coordinate

   Review the LEED section at the USGBC web page and review each LEED certification program and its objectives. For example, LEED for Existing Buildings focuses on operations and maintenance of existing buildings.

16. Which of these Sections has the greatest number of possible points?

   A. Sustainable Sites
   B. Water Efficiency
   C. Energy and Atmosphere
   D. Materials and Resources
   E. Indoor Environmental Quality

   Answer: C

   LEED Guide Reference: Introduction

   Study Area: Coordinate

   Each section of the LEED rating system can contribute points towards certification:

Sustainable Sites	14 points
Water Efficiency	5 points
Energy and Atmosphere	17 points
Materials and Resources	13 points
Indoor Environmental Quality	15 points

17. Project teams interested in obtaining LEED Certification for their project must do which of the following?

   A. Upon completion of construction, evaluate whether project qualifies for LEED certification
   B. Check building codes to see if LEED compliance is required
   C. Register their project with the USGBC
   D. All of the above

   Answer: C

   LEED Guide Reference: http://www.usgbc.org

   Study Area: Implement

Refer to the Guide and the USGBC web site and review the specific activities for the LEED application process. The process starts during the design phase and building codes are only relevant if they are required for compliance.

18. Which of the following is true about having a LEED-accredited Professional on your project team?

    A. Is a prerequisite for Innovation and Design Process category
    B. Gets you 1 credit in Innovation and Design Process
    C. Is a requirement for the commissioning agent
    D. All of the above
    E. None of the above

    Answer: B

    LEED Guide Reference: Innovation in Design

    Study Area: Implement

    Refer to the Guide and review the Innovation in Design Section. Review the process used to apply for and document ID points other than the 1 point allowed for having a LEEDAP on the project team.

19. Which of the following is NOT true regarding MR Credits 1.1, 1.2, and 1.3, Building Reuse?

    A. Significantly reduce construction waste
    B. Lower infrastructure costs
    C. Minimize habitat disturbance associated with greenfield site
    D. Often provide better-performing building envelopes

    Answer: D

    LEED Guide Reference: MR Credit 1: Building Reuse

    Study Area: Credit Knowledge

    The primary intent of MR Credit 1: Building Reuse is to conserve resources and minimize environmental impacts. Using an old building envelope may not be as efficient, in terms of performance, than building a new one. Credit strategy recommends upgrading components that would improve energy and water efficiency such as windows, mechanical systems, and plumbing fixtures.

20. Which THREE of the following items are required to obtain a LEED certification for any building project? (Choose 3)

    A. Comply with the mandatory provisions of ASHRAE/IESNA Standard 90.1–2004
    B. Provide fundamental commissioning of the building energy system
    C. Comply with mandatory provisions of IESNA Energy Conservation LBNL–47493
    D. Provide Fundamental Refrigerant Management

    Answer: A, B, D

    LEED Guide Reference: Credits

    Study Area: Coordinate

Refer to the Guide and review the *required standards* for each credit. See Question 3 for a more detailed description. For example, IESNA Energy Conservation LBNL–47493 is used to measure energy consumption, but is not *required*.

21. An owner is designing a 250-unit condominium project and plans to register for LEED Certification. The design team intends to apply for Sustainable Sites Credit 4.2: Alternative Transportation, Bicycle Storage, and Changing Rooms. The owner expects to have 550 occupants and plans to provide secure bicycle spaces including bicycle racks, lockers, and storage rooms that are easily accessible by building occupants during all periods of the year, free of charge. How many secure bicycle spaces must the owner provide in order to obtain 1 point towards LEED Certification?

    A. 28
    B. 55
    C. 83
    D. None of the above

    Answer: C

    LEED Guide Reference: SS Credit 4.2: Alternative Transportation, Bicycle Storage and Changing Rooms

    Study Area: Verify

    Refer to the Guide and review the specific requirements and calculations for each section. Sustainable Sites Credit 4.2: Alternative Transportation, Bicycle Storage and Changing Rooms requires facilities to be installed for *15% of building occupants* if the project is residential (condominium). If a building will have 550 occupants then 83 bicycle spaces will be necessary to achieve a point (*550 occupants * 15% = 83 spaces*)

22. Which THREE of the following items are required to obtain a LEED certification for a new building? (Choose 3)

    A. Storage and Collection of Recyclables
    B. Controllability of Systems
    C. ETS Control
    D. Construction Activity Pollution Prevention

    Answer: A, C, D

    LEED Guide Reference: EA Credit 6: Controllability of Systems

    Study Area: Implement

    Refer to the Guide and review the definition of a prerequisite and the requirement for each credit. For example, EA Credit 6: Controllability of Systems may be used to achieve a point but since it is not a prerequisite it is not *required*.

23. Your project intends to reuse a significant portion of the building shell, structure, and non-shell components. Which of the following statements is NOT true?

    A. You may be able to claim up to three credits for building reuse
    B. Building materials under this credit may be considered salvage materials

C. Preserved materials also may be applied towards Construction Waste Management credit

D. All of the above

E. None of the above

Answer: B

LEED Guide Reference: MR Credit 3: Material Reuse

Study Area: Implement

Refer to the Guide and review the *requirements* for each credit. MR Credit 3: Material Reuse only allows credit for fixed items that no longer serve their original purpose. MR Credit 1.1: Building Reuse allows points for

• Maintaining 75% of Existing Walls, Floors & Roof (1 point).
• Using 95% of Existing Walls, Floors & Roof (1 point).
• Using 50% of Interior Nonstructural Elements (1 point).

24. An owner is designing a four-story office building that will be registered for a LEED Certification. The project team has estimated that the total cost of materials for the project (CSI Divisions 2–10) will be $13,587,000. The owner intends to install bamboo flooring that will cost $165,000, cotton batt insulation at $21,000, and wool carpeting at $95,000. Which of the following items will allow for 1 point for MR Credit 6: Rapidly Renewable Materials?

A. Recycled brick veneer at $145,000

B. Linoleum flooring at $65,000

C. Wheatboard cabinetry at $55,000

D. Cork flooring at $53,000

Answer: B

LEED Guide Reference: MR Credit 6: Rapidly Renewable Materials

Study Area: Credit Knowledge

MR Credit 6: Rapidly Renewable Materials requires the use of rapidly renewable building materials and products (made from plants that are typically harvested within a 10-year cycle or shorter) for 2.5% of the total value of all building materials and products used in the project, on the basis of cost.

25. Which TWO of the following will achieve points towards a project's LEED Certification? (Choose 2)

A. Exceeding the standards set in the LEED Credit requirements

B. Involving a LEED Accredited professional to guide the project

C. Fundamental Refrigerant Management

D. Verifying installation, functional performance, and operation maintenance documentation.

Answer: A, B

LEED Guide Reference: EA Prerequisite 3: Fundamental Refrigerant Management

Study Area: Coordinate

Refer to the Guide and review the definition of a prerequisite and the requirement for each credit. Fundamental Refrigerant Management and verifying installation, functional performance, training, and operation maintenance documentation are prerequisites and don't contribute towards points.

26. As a general rule of thumb, the Innovation in Design credit for exceptional performance is awarded for which TWO items? (Choose 2)

A. Doubling the credit requirements
B. Exceeding the credit requirement
C. Achieving the next incremental percentage threshold.
D. Complying with a higher standard of measure

Answer: A, C

LEED Guide Reference: ID Credit Section

Study Area: Implement

Refer to the Guide and review the *requirements* in the Innovation in Design Section. The measurement of *exceptional performance* must be quantifiable.

27. The design team on a LEED Registered project is pursuing a unique strategy for improving ventilation effectiveness. It is unclear, on the basis of information provided in the Rating System and Reference Guide, whether or not the strategy satisfies the intent and requirements of a particular LEED credit and thus whether it might allow the project to earn much-needed points. After the team has read the Reference Guide with care, what is the next most important step the team should take toward determining the suitability of their design strategy for achieving the credit in question?

A. Search the list of LEED Certified projects for those having previously achieved the same credit
B. Send an inquiry to the TAG committee for the associated LEED credit category requesting clarification
C. Go to LEED-Online and complete a Credit Interpretation Request asking for clarification
D. Search the existing Credit Interpretation Rulings to see if there is one that addresses their question

Answer: D

LEED Guide Reference: http://www.usgbc.org

Study Area: Implement

Refer to the Guide and the USGBC web page and review the Credit Interpretation Rulings process and requirements. Once a project is registered, you have access to previous Credit Interpretation Rulings, and this is the first place to look when you are evaluating credits or design strategies.

28. The use of modulating photoelectric daylight sensors applies to which of the following areas?

A. Optimize Energy Performance
B. Daylight and Views
C. Light Pollution Reduction

    D. Measurement and Verification

    E. All of the above

    F. None of the above

Answer: A

LEED Guide Reference: EA Credit 1: Optimize Energy Performance

Study Area: Credit Knowledge

Refer to the Guide and review the *technologies* used for each of the credits. For example *daylight sensors* are used to turn a building's lights off when enough natural light is available, in order to decrease energy use (i.e., optimize energy performance).

29. A design team is about to submit its project's LEED registration to the USGBC, but now some members of the team are questioning a proposed strategy for an ID Credit 1–1.4, Innovation in Design. Which of the following aspects should be considered as the team evaluates this strategy?

    A. Is there a measurable environmental benefit to this strategy?

    B. Is this an existing ID credit that has been previously accepted by the USGBC?

    C. Is this strategy already covered by an existing LEED Credit category?

    D. Is there a Credit Interpretation Request?

    E. All of the above

    F. None of the above

Answer: E

LEED Guide Reference: http://www.usgbc.org

Study Area: Implement

Refer to the Guide and the USGBC web page and review the Credit Interpretation Rulings process and requirements.

30. Your client's project will reuse an existing building shell, keeping the entire roof, floor, and all exterior walls; reduce the number of existing parking spaces to create an open space equal to 50% of the site area (excluding building footprint); and use native plants and adapted and low-water landscaping. This strategy will help achieve which THREE of the following credits? (Choose 3)

    A. SS Credit 5.1: Site Development

    B. MR Credit 1.1: Building Reuse

    C. WE Credit 1.1: Water Efficient Landscaping

    D. MR Credit 1.3: Building Reuse

    E. SS Credit 7.2: Heat Island Effect

Answer: A, B, C

LEED Guide Reference: SS Credit 5.1: Site Development

Study Area: Coordinate

Refer to the Guide and review the requirements for each credit. For example, SS Credit 5.1: Site Development is given for creating an open space equal to 50% of the site area (excluding building footprint).

31. Your new office building incorporates the use of modulating photoelectric daylight sensors and has installed PV cells that supply at least 2.5% of the building's total project annual energy cost. These strategies may contribute to a LEED Platinum rating by contributing to which TWO credits? (Choose 2)

    A. Optimize Energy Performance
    B. On-Site Renewable Energy
    C. Measurement and Verification
    D. Resource Reuse
    E. Green Power

Answer: A, B

LEED Guide Reference: EA Credit 1: Optimize Energy Performance

Study Area: Coordinate

Refer to the Guide and review the requirements and technologies used for each credit. For example, modulating photoelectric daylight sensors may help achieve credit for EA Credit 1: Optimize Energy Performance.

32. The MEP Consultant has designed a system that will include a carbon dioxide monitoring system, mechanical ventilation, individual controls for temperature and humidity, and a permanent monitoring system. This strategy will help achieve at least three credits. Which TWO reference standards must be used to show compliance? (Choose 2)

    A. ASHRAE 62.1–2004, Appendix C
    B. ASHRAE/IESNA 90.1–1999
    C. ASHRAE 55–2004
    D. The Energy Policy Act (EPAct) of 1992
    E. ASHRAE 129–97, Air-Change Effectiveness

Answer: A, C

LEED Guide Reference: Credits

Study Area: Coordinate

Refer to the Guide and review the required standards for each credit (refer to Question 3 for definition of required standard.) For example, ASHRAE 62.1-2004, Appendix C, is used to evaluate EQ Credit 1: Outdoor Air Delivery Monitoring and EQ Credit 2: Increased Ventilation.

33. You must perform a Soil/Climate analysis to determine appropriate landscape types for compliance with which of the following areas?

    A. Water Efficient Landscaping
    B. Water Use Reduction
    C. Landscape and Exterior Design to reduce heat island effect
    D. Reduced Site Disturbance

E. All of the above

F. None of the above

Answer: A

LEED Guide Reference: WE Credit 1: Water Efficient Landscaping

Study Area: Coordinate

Refer to the Guide and review the process and requirements for each credit. WE Credit 1: Water Efficient Landscaping requires a soil and climate analysis to determine appropriate landscape types for compliance.

34. Strategies for reusing stormwater or greywater may be applied towards credit for which of the following areas?

   A. Water Use Reduction

   B. Water Efficient Landscaping

   C. Innovative Wastewater Technologies

   D. All of the above

   E. None of the above

   Answer: D

   LEED Guide Reference: Credits

   Study Area: Credit Knowledge

   Refer to the Guide and review the *requirements* and *definitions* for each credit.

35. The use of porous pavement systems with a high percentage of postindustrial waste applies to which of the following areas?

   A. Recycled Content

   B. Landscape & Exterior Design to reduce heat island effect

   C. Erosion and Sedimentation Control

   D. Resource Reuse

   E. Water Use Reduction

   Answer: A

   LEED Guide Reference: SS Credit 6.1: Stormwater Design

   Study Area: Credit Knowledge

   Refer to the Guide and review the technologies associated with each credit. Porous pavement may be used for reducing runoff and increasing infiltration for SS Credit 6.1: Stormwater Design, but it doesn't reduce the amount of water used.

36. Credit Interpretation Request rulings provide which TWO of the following? (Choose 2)

   A. Responses to written requests for interpretation of credit requirements

   B. Determination of whether a particular strategy can be used to satisfy two different credits at once

   C. Precedents for interpretation of LEED credits regarding specific strategies and applications

D. Definitive assurance that a particular method or strategy permitted on a previous project will be applicable to other projects in the future

Answer: A, C

LEED Guide Reference: http://www.usgbc.org

Study Area: Implement

Refer to the Guide and the USGBC web page and review the Credit Interpretation Rulings process and requirements. The CIR process only allows one question for one credit at a time. Answers are project specific and may not always be applicable to future projects.

37. Which of the following items is intended to reduce impacts from transportation and support the local economy?

    A. Site Selection
    B. Local/Regional Materials
    C. Alternative Transportation
    D. Renewable Energy
    E. All of the above
    F. None of the above

Answer: B

LEED Guide Reference: MR Credit 5: Regional Materials

Study Area: Credit Knowledge

Refer to the Guide and review each Credit *intent*. For example, the Regional Materials Credit is intended to support local economies.

38. Building on sites already degraded, using native plants, designing a compact building, and preserving existing natural site amenities (strategies used together or separately) influence all of the following areas EXCEPT:

    A. Stormwater Management
    B. Reduced Site Disturbance
    C. Water Efficient Landscaping
    D. Site Selection

Answer: D

LEED Guide Reference: SS Credit 6: Stormwater Design

Study Area: Coordinate

Refer to the Guide and review each credit *intent*. Stormwater Management assumes that if the site is already degraded, permeable surfaces decrease the amount of storm water runoff. The Storm Water Management plan compares predevelopment discharge with postdevelopment rates.

39. Maintenance of relative humidity levels between 30% and 60% will qualify for which of the following areas?

    A. Thermal Comfort
    B. Controllability of Systems

    C. Increase Ventilation Effectiveness

    D. Indoor Chemical and Pollutant Source Control

    E. All of the above

    F. None of the above

Answer: A

LEED Guide Reference: EQ Credit 7.1: Thermal Comfort

Study Area: Credit Knowledge

Refer to the Guide and review the *requirements* for each credit. EQ Credit 7.1: Thermal Comfort is given for providing a comfortable thermal environment that supports the productivity and well-being of building occupants. Thermal comfort is based on air temperature, radiant temperature, humidity, and air speed.

40. Who is included in the FTE?

    A. Full-time employees, visitors

    B. Full-time occupants, part-time occupants

    C. Full-time employees, part-time employees

    D. Full-time employees, part-time employees, customers

Answer: B

LEED Guide Reference: SS Credit 4: Alternative Transportation

Study Area: Credit Knowledge

Alternative Transportation attempts to reduce pollution and impacts from vehicle use. *Calculated* Full Time Equivalent factors are used to evaluate the project's parking areas. Strategies include minimizing the areas for parking and encouraging multi-passenger transportation (carpools, vanpools).

41. What would be useful in achieving WE Credit 1: Water Efficient Landscaping and SS Credit 5: Site Development?

    A. Vegetative swale

    B. Pervious parking

    C. Vegetative roof surface

    D. Rainwater reuse

Answer: C

LEED Guide Reference: WE Credit 1: Water Efficient Landscaping
SS Credit 5: Site Development

Study Area: Credit Knowledge, Implement

Water-efficient landscaping eliminates potable water use for landscape irrigation and reduces the overall need to irrigate. Site development conserves and protects existing natural habitat and restores damaged areas. Roofs built with natural vegetation can replace existing natural habitat removed or damaged for the project. Vegetated roofs allow rainwater reuse. The plant's water use helps to decrease stormwater runoff while eliminating the need for potable water for irrigation.

42. What percent of potable water use must be reduced to achieve advanced WE Credit 3?

A. 20%
B. 30%
C. 50%
D. 75%

Answer: B

LEED Guide Reference: WE Credit 3: Water Use Reduction

Study Area: Credit Knowledge

Water use reduction maximizes the building's efficiency and reduces the burden on the municipal water supply and wastewater systems. The percentage of potable water use reduction is used to evaluate the building's water use efficiency compared with the efficiency of water use by a traditionally designed building.

43. An owner wants to develop a previously developed site with imperviousness of 22% and a runoff of 200 fps (10,000 gallons per year). The postdevelopment imperviousness is 88%. What is the allowable amount of runoff rate (fps) and discharge (gallons) in order to achieve SS Credit 6?

A. 200 fps and 10,000 gallons
B. 500 fps and 20,000 gallons
C. 1,000 fps and 50,000 gallons
D. 1,200 fps and 100,000 gallons

Answer: A

LEED Guide Reference: SS Credit: Storm Water Design

Study Area: Credit Knowledge

Well-designed stormwater systems limit the disruption of natural site hydrology. Strategies include reducing impervious cover, increasing on-site infiltration, and managing stormwater. The site imperviousness and runoff rates are used to calculate the design's impact on the natural hydrology and size of the rainwater harvesting storage system to mitigate the impacts.

44. A building has the following plumbing fixtures specified for bathrooms:

5 low-flow toilets	1.1 gpf
4 low-flow sinks	0.5 gpm

One of the building's occupants uses the toilet and sink three times a day each and washes her hands for 15 seconds each time. How many gallons of water does she use each day?

A. 1.6 gal
B. 3.7 gal
C. 6.2 gal
D. 8.4 gal

Answer: B

LEED Guide Reference: WE Credit 2: Innovative Wastewater Technologies

Study Area: Credit Knowledge

Innovative wastewater technologies reduce wastewater generation and potable water use, while recharging the local aquifer. Expected building water use from the operation of sinks and toilets is calculated to evaluate the design's performance. Potential strategies include water-conserving plumbing fixtures, the use of nonpotable water (rainwater reuse, recycled building water) for building fixtures, and the treatment of wastewater on-site for reuse or infiltration. Calculations assume that female occupants will use the toilet three times per day.

45. Which THREE of the following items must be included in the submission to register intent for a project's LEED-NC Certification? (Choose 3)

   A. Owner's name
   B. Advisor's name
   C. Name and location of job
   D. Name of subcontractors that are responsible for templates
   E. Certification goal of project

Answer: A, B, C

LEED Guide Reference: http://www.usgbc.org
LEED-NC Overview and Process

Study Area: Implement

Project teams that want to obtain LEED-NC certification must first register their intent to the USGBC. This registration process is explained at the USGBC web page and in the Guide.

46. Which TWO of the following projects would most likely choose LEED-NC when registering for USGBC Certification? (Choose 2)

   A. New 40,000-sf building
   B. Remodel of a 40,000-sf building's HVAC system
   C. Remodel of a 40,000-sf building along with significant HVAC replacement
   D. Tenant Improvement of small building

Answer: A, C

LEED Guide Reference: http://www.usgbc.org
LEED-NC Overview and Process

Study Area: Coordinate

The USGBC has several Certification programs for different types of buildings and projects. Each project must be evaluated to see which program is appropriate before the team registers for Certification. The *requirements* for each program are explained on the USGBC web page along with a description of the process for registration.

47. Which THREE of the following activities are requirements of the commissioning authority to achieve EA Credit 3: Enhanced Commissioning? (Choose 3)

   A. Conduct one design review of the OPR and BOD
   B. Review contractor submissions

C. Develop HVAC design calculations

D. Review building operation within 10 months after substantial completion

E. Install and test HVAC equipment

Answer: A, B, D

LEED Guide Reference: EA Credit 3: Enhanced Commissioning

Study Area: Coordinate

For EA Credit 3: Enhanced Commissioning, the Commissioning Agent is an independent authority that reviews the design and verifies performance for building systems. The requirements for this position are explained in the Guide.

48. An owner is reusing existing walls, floors, and roofs but not enough to achieve MR Credit 1: Maintain 75% of Existing Walls, Floors and Roofs. Which TWO of the following strategies may allow for potential points towards project certification? (Choose 2)

    A. MR Credit 1.1: Building Reuse
    B. MR Credit 5: Regional Materials
    C. MR Credit 6: Rapidly Renewable Materials
    D. MR Credit 2.1: Construction Waste Management
    E. MR Credit 7: Certified Wood

    Answer: A, D

    LEED Guide Reference: MR Credit 1.1: Maintain 75% of Existing Walls, Floors and Roofs

    Study Area: Implement

    Each credit section of the rating system has specific *requirements* and *intents* for achieving points that must be considered for all of the building systems holistically. For example, MR Credit 1.1: Building Reuse extends the life cycle of existing structures by reusing an existing building's structure and envelope.

49. Which of the following items is NOT a strategy for achieving a point for EA Credit 2: On-Site Renewable Energy?

    A. Photovoltaic panel system
    B. Solar thermal water heat
    C. Geothermal water source heat pump and generator
    D. Low-impact hydro turbine
    E. Biomass generator

    Answer: B

    LEED Guide Reference: EA Credit 2: On-Site Renewable Energy

    Study Area: Implement

    Each credit section of the rating system has specific *strategies* and *intents* for achieving points that must be considered for all of the building systems. The intent of EA Credit 2: On-Site Renewable Energy is to decrease energy use by generating power on the building site. Solar water heat decreases or eliminates the amount of energy used to heat a building's water by using natural sunlight to generate heat but does not produce electricity.

50. For a 1,000,000-sf office building (11.7 kWh/sf-yr, 58.5 kBtu/sf-yr) in New York ($0.1113/kWh, $0.00895/kBtu), determine how many dollars' worth of renewable energy is required to achieve 1 point for EA Credit 2.

    A. $45,600
    B. $57,800
    C. $69,150
    D. $71,560
    E. $93,750

Answer: A

LEED Guide Reference: EA Credit 2: On-Site Renewable Energy

Study Area: Credit Knowledge

Each credit section of the rating system has specific requirements for achieving points that must be considered for all of the building systems. The intent of EA Credit 2: On-Site Renewable Energy is to decrease energy use by generating electric power on the building site. The effectiveness of the system is evaluated by comparing its offset to building energy cost and requires a decrease of at least 2.5% in annual energy cost in order to be considered effective. Refer to LEED Guide for example calculation.

51. In order for a building to achieve points for EA Credit, its minimum energy performance must be established and compared against baseline calculations. Which of the following methods is used to calculate the unit cost value for baseline energy performance?

    A. Total average cost of energy divided by the total average annual expected energy consumption
    B. Annual expected energy consumption divided by the total average cost of energy
    C. Average monthly voltage drop from utility
    D. Energy Cost Budget Compliance Report

Answer: D

LEED Guide Reference: EA Prerequisite 2: Minimum Energy Performance
EA Credit 1: Optimize Energy Performance

Study Area: Verify

The intent is to establish the minimum acceptable level of energy efficiency for a proposed project. This is a prerequisite for achieving points for any EA credit and is set on the basis of compliance methods set forth in ASHRAE/IESNA Standard 90.1–2004. The energy performance for each of the building systems and for the building envelope are evaluated and compared against a baseline model. These represent the prerequisite energy efficiency. Projects are encouraged to *optimize* energy performance by offering up to 10 points for EA Credit 1.

52. Which TWO of the following Credits require decision making by the contractor, design team, and owner? (Choose 2)

    A. EA Prerequisite 1: Fundamental Commissioning of the Building Energy System
    B. EA Prerequisite 3: Fundamental Refrigerant Management

C. EA Credit 3: Enhanced Commissioning

D. EA Credit 5: Measurement and Verification

E. EA Credit 6: Green Power

Answer: A, C

LEED Guide Reference: EA section Overview: Table 1

Study Area: Coordinate

The decision-making process is an important part of successfully implementing the project goals. A decision matrix is a valuable tool for projects such as the EA section Overview: Table 1 in the LEED Guide.

53. Which of the following Credits is a construction phase submission?

A. EA Credit 2: On-Site Renewable Energy

B. EA Credit 4: Enhanced Refrigerant Management

C. EA Credit 6: Green Power

D. All of the above

E. None of the above

Answer: D

LEED Guide Reference: Introduction, LEED Green Building Rating System Review and Certification

Study Area: Implement

The LEED project registration includes a two-phase application process including a design and a construction or final submission. This process is outlined in the LEED Guide and on the web site. Each submittal includes specific Credits.

54. ASHRAE Standard 55–2004 relates to which of the following areas?

A. Thermal comfort conditions

B. Building ventilation

C. Individual thermal controls

D. Air-cleaning devices

Answer: A

LEED Guide Reference: EQ Credit 7.1 Thermal Comfort

Study Area: Coordinate

Projects should be designed to provide a comfortable thermal environment for building occupants that improve productivity and well-being. The ASHRAE Standard 55–2004: Thermal Comfort Conditions for Human Occupancy is a guideline for designing HVAC systems and building envelopes.

55. The Solar Reflectance Index (SRI) is a measurement of a material's ability to

A. Reject solar heat

B. Reflect solar light waves

C. Solar surface radiation

D. Absorb solar light waves

Answer: A

LEED Guide Reference: SS Credit 7.1: Heat Island Effect

Study Area: Verify

Heat island effect can harm microclimates and human and wildlife habitat. The Solar Reflectance Index is a measurement of the ability of a material (such as asphalt or roofing) to reject solar heat and is used to evaluate any resultant heat island effect.

56. For a two-phase application, which THREE of the following items are included in the design phase submission? (Choose 3)
   A. EQ Prerequisite 1: Minimum IAQ Performance
   B. EQ Credit 2: Increased Ventilation
   C. EQ Credit 4: Low-Emitting Materials
   D. EQ Credit 6: Controllability of Systems

Answer A, B, D

LEED Guide Reference: Introduction, LEED Green Building Rating System Review and Certification

Study Area: Implement

Each credit is designated if it is allowed in the initial design submission for projects that choose the two-phase application process.

57. The Glazing Factor Calculation is an option to evaluate EQ Credit 8.1: Daylight and Views. Which THREE of the following are used to calculate the Glazing Factor? (Choose 3)
   A. Floor area
   B. T_{vis}
   C. Window height factor
   D. Glass SRI
   E. Gross building area

Answer: A, B, C

LEED Guide Reference: EQ Credit 8: Daylight and Views

Study Area: Verify

Designs that incorporate natural daylighting and views provide a connection to the outdoors for building occupants. Glazing factors are calculated to evaluate how the design's use and location of windows relates to the interior space.

58. To satisfy EQ Credit 8: Daylight and Views, vision glazing must be placed between 2'-6" and 7'-6" above the finished floor for 90% of occupied areas. The square footage must meet which TWO of the following criteria? (Choose 2)
   A. In plan view, the area within sight lines from perimeter glazing
   B. In plan view, the area with direct sight lines to new or existing landscaping
   C. In section view, a direct sight line drawn from the area to perimeter vision glazing
   D. In section view, a direct sight line drawn from the area to a skylight (if interior wall obstruction only)

Answer: A, C

LEED Guide Reference: EQ Credit 8: Daylight and Views

Study Area: Verify

Window locations are an important design consideration and should be evaluated relative to the interior space to maximize daylighting and views.

59. ASHRAE/IESNA 90.1–2004 is a standard used to evaluate which of the following items?
    A. Energy efficiency
    B. Ventilation effectiveness
    C. Thermal comfort
    D. Refrigerant use

Answer: A

LEED Guide Reference: EA Prerequisite 2: Minimum Energy Performance

Study Area: Coordinate

ASHRAE/IESNA 90.1–2004 is a standard that is used to evaluate several of the LEED Credits.

60. A project needs to achieve more than 4 points for EA Credit 1: Optimize Energy Performance. Which of the following compliance path options would you recommend?
    A. Whole Building Energy Simulation
    B. Prescriptive Compliance Path–ASHRAE
    C. Prescriptive Compliance Path–Advanced Building Benchmark
    D. Energy Model–Test Case

Answer: A

LEED Guide Reference: EA Credit 1: Optimize Energy Performance

Study Area: Implement

Each Credit has specific *options* for documenting compliance with the requirements. The best option should be selected for each project that will help optimize energy performance.

61. Which THREE of the following activities must be completed by the project Commissioning Agent in order to achieve a point for EA Credit 3: Enhanced Commissioning? (Choose 3)
    A. Conduct at least one review of the OPR and BOD
    B. Review contractor submissions for systems being commissioned
    C. Develop systems manuals for future operating staff
    D. Verify that personnel training requirements are completed
    E. Review building operations within 10 months after substantial completion

Answer: A, B, E

LEED Guide Reference: EA Credit 3: Enhanced Commissioning

Study Area: Verify

To maximize a building's energy efficiency, the design should be based on an evaluation of each system. The system should be calibrated and monitored for energy efficiency. A Commissioning Agent can help evaluate the design and performance.

62. Which THREE of the following options may be used to achieve EQ Credit 8: Daylight and Views? (Choose 3)

    A. Daylight Simulation Model
    B. Glazing Factor Calculation
    C. Building Model
    D. Daylight Measurement

    Answer: A, B, D

    LEED Guide Reference: EQ Credit 8: Daylight and Views

    Study Area: Credit Knowledge

    Each Credit has *options* for documenting compliance with the requirements. These options will help projects evaluate performance in each of the LEED sections.

63. Which of the following criteria relates to thermal comfort for building occupants?

    A. Air temperature
    B. Radiant temperature
    C. Relative humidity
    D. All of the above
    E. None of the above

    Answer: D

    LEED Guide Reference: EQ Credit 6.2: Controllability of Systems

    Study Area: Verify

    The productivity, comfort, and well-being of building occupants are affected by the indoor environment. Well-designed buildings provide systems that allow occupants to adjust such things as ventilation, heat, air-conditioning, and task lighting to individual preferences.

64. Which THREE of the following items are required to comply with EQ Credit 5: Indoor Chemical and Pollutant Source Control for a mechanically ventilated building? (Choose 3)

    A. Provide 6-foot walk-off mat
    B. Create negative pressure in areas with chemicals present
    C. Test within 10 months after occupancy
    D. Install MERV-13 filters

    Answer: A, B, D

    LEED Guide Reference: EQ Credit 5: Indoor Chemical and Pollutant Source Control

    Study Area: Credit Knowledge

    Each Credit has specific *requirements* that must be documented and registered in order to achieve a point. EQ Credit 5: Indoor Chemical and Pollutant Source Control is intended to minimize and control pollutant entry into buildings and cross contamination of building spaces. This Credit requires a minimum six-foot walk-off mat at each building entry to capture dirt before it enters the building.

65. Renewable energy sources used to achieve EA Credit 6: Green Power are defined by which of the following organizations?

    A. Center for Resource Solutions

    B. Department of Energy

    C. Renewable Energy Institute

    D. ASHRAE

Answer: A

LEED Guide Reference: EA Credit 6: Green Power

Study Area: Coordinate

The Center for Resource Solutions (CRS) is an organization that evaluates the environmental impacts for different types of renewable energy sources. The CRS Green-e products certification can be used to evaluate each project. Review the *Requirements* for each Credit and *compliance standards*.

66. The project manager has just been informed that the carpet that was supposed to be delivered to the project is no longer available. What should the advisor do in order to make sure the new carpet is compliant and no points are lost?

    A. Test carpet for VOC emission

    B. Verify nonfriable content

    C. Verify Green Label Plus Program certification

    D. Evaluate expected useful life

    E. Recalculate the CRI factor

Answer: C

LEED Guide Reference: EQ Credit 4.3: Low-Emitting Materials

Study Area: Implement, Coordinate

Carpet systems should be evaluated to minimize indoor air contaminants. Carpets and glues contain volatile organic compounds (VOC) that are harmful and irritating to building occupants. Designers can evaluate carpet selection using the Carpet and Rug Institute's Green Label Program. Each Credit has specific *requirements* and *reference standards* used to verify compliance.

67. A project design specifies doors made with a composite agrifiber core and wood veneer. Toward what Credit will this product most likely contribute?

    A. EQ Credit 4.4: Low-Emitting Materials: Composite Wood and Agrifiber Products

    B. ID Credit: Innovation in Design

    C. MR Credit 3: Materials Reuse

    D. MR Credit 6: Rapidly Renewable Materials

    E. MR Credit 7: Certified Wood

Answer: D

LEED Guide Reference: MR Credit 6: Rapidly Renewable Materials

Study Area: Verify

Rapidly renewable materials are made from plants that can be harvested at least every 10 years, like bamboo, cotton, linoleum, cork, wool, and wheat. Rapidly renewable materials can be used for items such as insulation, flooring, doors, and cabinets (to name a few). Review the *strategies* for each Credit.

68. A project is trying to achieve the Rapidly Renewable Materials Credit with an overall cost of materials of $1,000,000. The project has already received shipments of $21,000 in renewable materials. Which item can be used in order to achieve one point?

A. Cotton insulation for $4,500
B. Agrifiber for $1,500
C. Bamboo flooring for $2,000
D. Linoleum flooring for $6,000

Answer: D

LEED Guide Reference: MR Credit 6: Rapidly Renewable Materials

Study Area: Credit Knowledge

Rapidly renewable material products are made from plants that are harvested within a 10-year cycle. Rapidly renewable materials such as cotton, agrifiber, bamboo, and linoleum have many commercial building applications. Using rapidly renewable material products helps to reduce the use and depletion of raw and long-cycle renewable materials.

69. Which TWO of the following items should an administrator consider when he or she evaluates a previously developed site for redevelopment? (Choose 2)

A. Located within $1/2$-mile radius from a residential area
B. Community density
C. Elevation relative to 100-year flood
D. Distance to closest body of water

Answer: A, B

LEED Guide Reference: SS Credit 2: Development Density and Community Connectivity

Study Area: Coordinate

When considering a previously developed site for redevelopment, an administrator should take into account the density and connectivity of the location in order to protect greenfields and preserve habitat and natural resources. It is best to develop in urban areas with existing infrastructure that provides occupants with a connection to the community and to essential services.

70. In order for a building to achieve SS Credit 4.2: Alternate Transportation, which of the following items must a commercial or institutional building provide?

A. Bicycle storage for at least 5% of building occupants
B. Covered bicycle storage for 15% of building occupants
C. Shower facilities for at least 5% of FTE occupants
D. All of the above
E. None of the above

Answer: A

LEED Guide Reference: SS Credit 4.2: Alternate Transportation

Study Area: Credit Knowledge

Designing bicycle storage and changing rooms in or near a project encourages occupants to ride their bikes to work instead of driving. The intent is to help reduce pollution and land development impacts from automobile use.

71. A project requires 2 tons of concrete which is made up of 100 lbs of fly ash. What percent of the fly ash can apply to MR Credit 4: Recycled Content?

   A. 5%
   B. 10%
   C. 15% preconsumer/25% postconsumer
   D. 25% preconsumer/15% postconsumer

Answer: A

LEED Guide Reference: MR Credit 4: Recycled Content

Study Area: Credit Knowledge

Using building products that are made from recycled materials reduces the impacts of extracting and processing virgin materials. Fly ash is recycled waste from coal burning and can be used in concrete to reduce water use.

72. EQ Credit 4.4: Low-Emitting Materials does not allow the use of urea-formaldehyde resins and applies to which TWO of the following items? (Choose 2)

   A. Interior doors
   B. Furniture
   C. Concrete slab formwork
   D. Composite wood floor

Answer: A, D

LEED Guide Reference: EQ Credit 4.4: Low Emitting Materials

Study Area: Credit Knowledge

The resins used for wood fabrication can be harmful and affect indoor air quality. Resins used for items such as composite wood, agrifiber products, and adhesives should contain no added urea-formaldehyde.

73. Which of the following is NOT a likely strategy for achieving SS Prerequisite 1: Construction Activity Pollution Prevention?

   A. Temporary seeding
   B. Mulching
   C. Earth dike installation
   D. Sediment basin installation
   E. All of the above
   F. None of the above

Answer: F

LEED Guide Reference: SS Prerequisite 1: Construction Activity Pollution Prevention

Study Area: Credit Knowledge

A construction Activity Pollution Prevention plan is used to reduce pollution from construction activities by controlling soil erosion, waterway sedimentation, and airborne dust generation. Review strategies for achieving each credit requirement.

74. Which TWO items will most likely contribute towards credit for MR Credit 2: Construction Waste Management? (Choose 2)

    A. Foundation excavation soil that can't be reused on-site

    B. Excess paint

    C. Demolished copper water pipe

    D. Land-clearing debris

    E. Lumber scraps

Answer: C, E

LEED Guide Reference: MR Credit 2: Construction Waste Management

Study Area: Implement

Decreasing and managing the waste on a construction project can reduce the amount of materials used and waste generated. The intent is to reduce emissions from transporting waste and to limit the impacts on landfills. Not included are items such as hazardous waste, soil, and debris from land clearing.

75. Before submitting a Credit Interpretation Request (CIR), the applicant must

    A. Review the intent of the credit or prerequisite in question and self-evaluate whether the project meets this intent

    B. Consult the LEED Reference Guide for more detailed explanation, instructions, calculations, and guidance

    C. Review the CIR pages to see whether the same inquiry has been answered previously or whether there are relevant CIRs that can help him or her deduce the answer

    D. Contact LEED customer service to look into the request and confirm that it warrants a new CIR

    E. All of the above

Answer: E

LEED Guide Reference: http://www.usgbc.org

Study Area: Implement

Refer to the Guide and the USGBC web page and review the Credit Interpretation Request process and requirements. Once a project is registered, applicants have access to the database of previous CIR Rulings, which is the first place to look when evaluating credits or design strategies.

76. A project intends to use the flush-out compliance path to achieve EQ Credit 3.2: Construction IAQ Management Plan. Which of the following activities can be done concurrently with the flush-out?

    A. Countertop installation using low-VOC glue

    B. Final cleaning

    C. Final HVAC testing and balancing

D. All of the above

E. None of the above

Answer: E

LEED Guide Reference: EQ Credit 3.2: Construction IAQ Management Plan

Study Area: Implement

Building flush-out before occupancy improves indoor air quality. Construction activities should be complete before flush-out in order to avoid contamination.

77. Which of the following project team members is most likely to carry decision-making responsibility for SS Credit 5.1: Site Development, Protect or Restore Habitat?

A. Contractor

B. Owner

C. Design team

D. All of the above

E. None of the above

Answer: D

LEED Guide Reference: SS Credits: Sustainable Sites

Study Area: Coordinate

Involving the right team members in the decision making process is critical for achieving success. During the site development phase, the contractor, design team, and owner collaborate to ensure the protection and restoration of wildlife habitat.

78. What is the general rule of thumb for achieving ID Credits for exemplary performance? (Choose 2)

A. Doubling the credit requirements

B. Achieving the next incremental percentage threshold

C. Increasing performance by a minimum of 50%

D. Exceeding baseline calculations by 50%

Answer: A, B

LEED Guide Reference: ID Credit 1: Innovation in Design

Study Area: Coordinate

Innovative designs are encouraged and can be used to achieve points for exceptional performance. Each strategy is evaluated on the basis of its merit.

79. The contractor is responsible for documenting which TWO of the following items during the project's construction phase? (Choose 2)

A. Provide MSDSs with paint submissions

B. Provide energy modeling calculations

C. Provide site photometric plans

D. Provide calculations for waste diverted from landfill

Answer: A, D

LEED Guide Reference: Credits

Study Area: Implement

It is important to understand the roles and responsibilities of each team member in order to implement and achieve the project goals.

80. The owner of a 50-year-old office building has decided to replace the mechanical system to improve energy efficiency and is implementing green procurement and operations policies. Which LEED rating system is most applicable to this project?

    A. LEED-CI
    B. LEED-EB
    C. LEED-CS
    D. LEED-NC
    E. LEED-ND

Answer: B

LEED Guide Reference: Introduction

Study Area: Implement

It is important to understand strategies for different types of building project types in order to achieve the project goals. The USGBC has different rating systems that apply to different types of projects.

INDEX